The R.A.M.S. Library of Alchemy

Volume 11

The Greater and Lesser Edifyer

by Johann Grashoff

Translated by

Rob Firmage

R.A.M.S. Publishing Company

The Greater and Lesser Edifyer

by Johann Grashoff

Translated by

Rob Firmage

Produced by

Restorers of Alchemical Manuscripts Society
1984

R.A.M.S. Publishing Company

R.A.M.S. Publishing Company
117 Rutherford Lane
Stuarts Draft VA 24477

The Greater and Lesser Edifyer

First Edition 2015

ISBN-13 **978-1508984870**
ISBN-10 **1508984875**

Image Processing by Philip N. Wheeler

This book is sold for informational purposes only. Neither the
publisher nor the editor shall be held accountable for the use or misuse
of the information in this book.

Printed in the United States of America

Table of Contents

Dedicated to Hans W. Nintzel,
American Alchemist
and
Founder of the
Restorers of Alchemical Manuscripts Society
(R.A.M.S.)

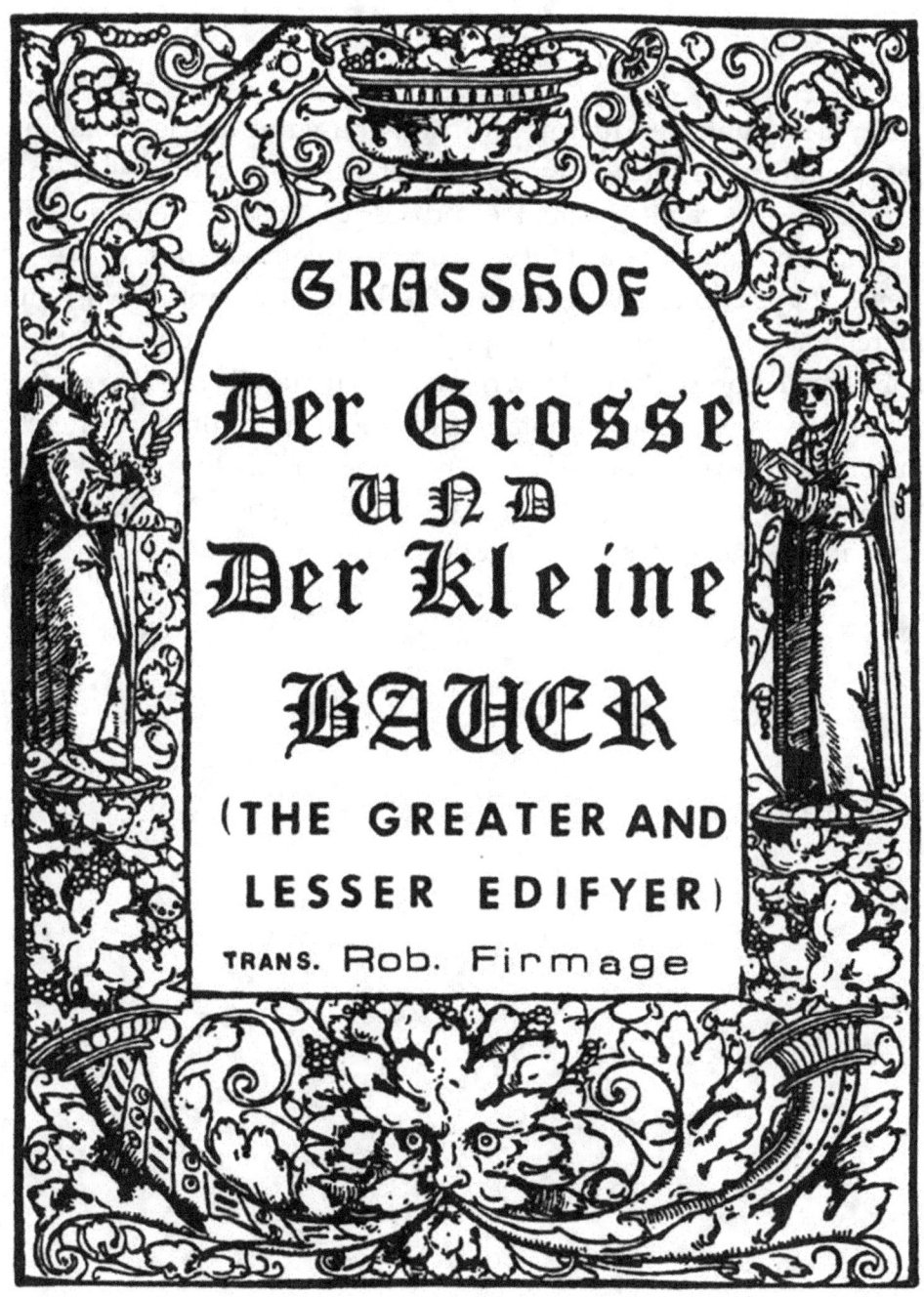

GRASSHOF

Der Grosse
und
Der Kleine
BAUER

(THE GREATER AND
LESSER EDIFYER)

TRANS. Rob. Firmage

Disclaimer

Liability: The publisher does not warrant or assume any legal liability or responsibility for the accuracy, completeness, or usefulness of any information, apparatus, product, or process disclosed. The publisher makes no representation as to the accuracy or completeness of the contents of this book and specifically disclaims any implied warranty of merchantability or fitness for a particular purpose. No warranty may be created or extended by written sales materials or sales representatives. You should obtain professional consultation where appropriate. The publisher shall not be liable for any loss of profit or other commercial or personal damages, including but not limited to special, incidental, consequential, or other damages.

APERTA   ARCA   ARCANI
artificiossimi

or

THE REVEALED AND OPEN CHEST
OF THE GREATEST OF ALL AND MOST ARTIFICIAL

secrets of nature

OF

the GREATER and the LESSER edifyer

TOGETHER WITH THE PROPER AND TRUE

PHYSICA NATURALI ROTUNDA
Described comprehensively through a

VISIONEM CHYMICAM QABALISTICAM

A WARNING, INSTRUCTION AND DEMONSTRATION

Against all those who falsely persuade others to
prepare the **Aurum Potabile** without the tincture
of the **Universal Lapidis Philosophici,** *per se*
and in a little time.

Hamburg   and   Stockholm

Published by Gottfried Liebezeit, Bookseller
Printed by Abraham Aubrem, the Counts Printer
Hanauis, Hoffbuch Druckerei, anno MDCLXXXVII

IN SUPERIORI EST IN MEDICO
FONTIS VENA, QUA EST PHILO-
SOPHORUM REGULA PRIMA.

HERMES.

METALLA OMNIA QUOAD PURI-
TATEM SUNT EJUSDEM SUB-
STANTIAE: DIFFERUNT TAN-
TUM IMPURITATE, MAJORI
VEL MINORI DIGESTIONE.

ARROS.

Introduction

Philip N. Wheeler

Johann Grasshoff (or Grasshof, Grasse) (1560 – 1623) was a Pomeranian jurist and alchemical writer. He is also recorded as a medical advisor to Ernest of Bavaria, an Episcopal counselor.

His writings include the *Aperta Arca arcani artificiosissimi* (1617) and a *Cabala Chymica* (1658).

The compilation of the 1625 *Dyas chymica tripartita* is also attributed to him; it includes *The Golden Age Restored* of Henricus Madathanus, *The Book of Lambspring* of Nicholas Barnaud, and the *Book of Alze*.

PREFACE

Although I have never been engaged in Alchemical matters, because, as you should know, this is not my occupation; nevertheless I have taken no little delight and pleasure in a proper recognition of the true powers of the creatures of Almighty GOD. For, in so far as I at times from need, but also occasionally from desire have read through a philosophical book, I have noted at least this much from it, that something more than the four elements are contained in everything, and that these are only a receptacle, or a shell, of an indwelling, divine, solely efficacious spirit: a fifth essence. And that beyond this, there lie hidden in the Lapide Philosophorum, copious and incredible secrets, which are beyond all human understanding. Thus, it is very easy to set forth, to convince, to demonstrate, and also, as it were, to point with one's finger toward the true certainty of the praiseworthy order of the R. C.; For, as is known to GOD-fearing hearts, such exalted secrets are contained in nature (of which I will mention only a few) so why then, if such is in the power of Christian love, should they not have and practice a beatific Brotherhood among themselves? M. Petrus Bonus Lombardus reminds us in his Margarita Preciosa Novella, chapt. 6., and D. Augustinus testifies also in Sumrna Confessionum (as well as many other highly enlightened and beatific men) that the wise Plato, who stands nearer the saving faith than all other heathens, and who lived more than three hundred years before the birth of our Savior and Messiah Jesus Christ, is described in the Holy Gospel of the Holy Evangelist and Apostle John (in the verses beginning "In the beginning was the Word," up to the passage "There was a man sent from GOD.."). That this science has arisen from Plato, and Saint John should have kept his FORMULA so cryptic, should be well contemplated by everyone. The above-mentioned MARGARITA ascribes to him the LAPIDEM PHILOSOPHORUM, and, in my opinion, not improperly, as can be inferred with certainty from the following points, as well as from other books.

In The Revelations of the Secrets of Alchemy, divulged by M. Henricum Vogelium, Parson of Lutzelstein, you will find in Book 4, chapter 7 that the heathens had recognized the Creator through his Creation, and had seen an image of the Holy Trinity (three different persons) in one single inseparable being (as Lombardus also attests in chapter 13) and that this is not difficult to demonstrate in common alchemical work. He also reports that, from the light of Nature, the heathens had recognized that the Son of God would become a man, that a Virgin would be his mother, and that Jesus Christ would have two births: One from eternity and one in time, as well as other points concerning his person. For instance, that man is the little world. That original sin is innate in him, and that there is only one intermediary between GOD and man, who redeems the human race through his suffering and death. That the power and activity of man is nothing, but that GOD'S word is powerful and healing. That GOD works powerfully through intermediaries, that good works follow unforced and from themselves, and that man may partake of the holy body and blood of Jesus Christ, and thus be born, die and be resurrected. That GOD will create a new heaven and earth, and so on. You will find much about the visible appearance of the creation of the heaven, earth and everything within them, as well as innumerable other very exalted secrets in this book. Read, in addition to this, HARMONIA LUMINIS, GRATIAE ET NATURAE; the FAMAM AND CONFESSION of the highly enlightened Brotherhood of the R. C. M.; Valentinum Wergelium; Philippus Mornaeus DE VERITATE RELIGIONIS CHRISTIANAE; Mutius Pansa's DE OSCULO ETHNICAE ET CHRISTIANAE RELIGIONIS; Franciscus Tidicae's MICROCOSM; Alexander von Suehten; etc. And although these partial articles of faith and knowledge may be

clearly seen through the revealed word of GOD, the
Book of Grace; nevertheless, Almighty GOD has also
written such images in the Book of Nature, that is
in heaven and earth and everything that is within
them, in order that these two books may agree with
one another, as in the light of day. For how could
Nature, the most beautiful creation, be repugnant to
GOD, since the Creator took so much pleasure in it,
and worked on it daily in accord with his delight?
It is thus quite possible for the heathens to have
attained to the Book of Grace by means of this
natural Book, as, beyond all doubt, the Philosophi
Magi or wise men from the East were led to Bethlehem
first by means of the natural light, and thereafter
by the supernatural appearance of the Star. Why then
should anyone still doubt concerning that to which
the blessed brothers of the R. C. have attested, and
with which the most wise King Soloman and ancient
wise men as well as men as contemporary as the Echo
of the Fraternity, Julianus de Campis, and others
agree?

Beyond this, you may consider what the Urim and
Thummim were, in the breast-plate of Aaron the
Highpriest in the Wonderbook of GOD, that is, the
Holy Bible, Exodus 28. For instance, how the holy
priests asked questions to the Lord by means of them
or how in 1 Kings 23 and 30, David questioned the

Lord by means of the URIM; or in chapter 28, how
Saul complained that the Lord did not answer him in
dreams, nor through the URIM, nor through prophets.
Note also how Moses burned the incombustible gold in
Exodus 32. How this came from a little powder of
earth (4 Esd. 8) and how it turned translucent like
glass in Revelations 21. Think how the fire turned
to a thick water in 2 Macc. 1. Consider whether the
holy blood of Jesus Christ, which was spilled upon
the cursed earth has not the power again to make it
holy. Think also how the creatures are subject to
vanity against their will, and long for, sigh for
and look forward to their end (Romans 8). But this
is enough for now; do not be a dumb pillar of salt
or of a stubborn heart. Nevertheless, in regard to
this tractatus the first two parts are written by a
Doctor of Law, as is mentioned in them. You should,
incidentally, perceive that they should be entitled
the Edifyers (the Greater and the Lesser - this
LILIUM INTER SPINAS). The other part has indeed been
published previously, but very falsely, incorrectly
and reduced by half. You can find it In Conferendo,
whose author is called Fransiscus Grellius in a
Latin copy. Thus, my favoured reader, may you be
satisfied with this for now, and if I should find
that my industry pleases you, I will then endeavour
that even more wonderful things of this sort should

see the light of day. With this I commend you to the grace of GOD.

FIRST PART

THE OPEN CHEST OF THE GREATEST SECRETS OF NATURE BEGINNING WITH THE GREATER EDIFYER (OR: FARMER)

Prosperity from Him who is the
beginning and the end.

The holy Apostle James says in Chapter 1: "Every good gift and every perfect gift is from above, and cometh down from the Father of lights, with whom there is no variableness, neither shadow of turning."

St. Paul says in 1 Corinthians 3: "I have planted, Apollo watered; but GOD gave the increase."

Alanus Phil: "Son, set your heart more on GOD than on the Art, for it is but a gift from GOD, and He imparts it unto whomsoever He will; therefore, have peace and joy in GOD, and you shall have the Art."

Alphidus: "Son you must know that you cannot attain to this Art, unless you have purified your heart and your mind toward GOD, and have sincerely reconciled yourself with Him; for as soon as GOD sees that you are of an upright and well meaning disposition, then through His grace, He shall allow you to master this Art."

Dionys. Zach. (Fol. 69): "For you it is meet that you read the philosophers with unflagging patience and constancy before you set your hand to your work, asking GOD at all times for understanding through His grace; for no one attains to this Art through luck or through chance, but rather more through prayer than through other means, though other means must be also employed."

The great King and philosopher Hermes (in Lib. di Nuitate Entis); "Direct yourself fully and completely to Him who is above you, lift the wings of your understanding to the radiance of the higher substance, for then you shall perceive with your eyes, both inwardly and outwardly, its countless and exalted beauties, and another light, which surpasses all light. You shall be astonished and give heed no longer to all the works of the world, and in your heart shall also wish and choose a premature death, shall castigate and mortify your body, and beyond this, shall even deny and hate your own soul. You shall honour, the King of Kings and the GOD of glory with heartfelt and beautiful hymns of praise. For we should worship and love with heart and soul the Word of the Father, who has so loved us and is the heart of the Father; and be astonished at so great a treasure, and glorify Him, that we may finally be worthy of the community of GOD, and be fulfilled by Him through His grace, to whom be all praise and honour from Eternity to Eternity. Amen.

I have set these testimonies or witnesses in this place, in order that no one should take it into his head to desire to achieve this Art by himself alone (let it be by whatever means one might please). No; for it is the will of GOD that all things should be prayed for; and especially this, since it is the highest thing next to the human soul that GOD has created. It is a mirror of all the highest and lowest things, in which GOD'S essence may be seen sufficiently, as in a mirror. It is the Mercurius Vitae without which no man, animal or plant may live (in this regard read Clangor Bucc., fol. 475 and PRINC. Fol. 468 in fine, fol. 474 in fine). Therefore my dear friend and brother in Jesus Christ, whoever you may be, and to whom I entrust this secret out of Christian love, keep GOD before your eyes, pray, read the philosophers and not the writings of the sophists, work with patience, and you shall behold miracles. And should you attain to it, do not misuse it; or the punishment of GOD shall follow upon your heels. Keep it a Secret; I have entrusted it to you for your own well-being, and you shall be held accountable for it on the Day of Judgment. Amen.

It is written by the wise and mighty King GEBER. (c. 1. part 1 Summa Perfect) that whoever does not understand the basic fundamentals and causes of the growth of metals is already widely separated from our ART, for he has no well grounded roots, upon which to found and establish the certainty of his undertaking. This is confirmed also by the excellent philosopher Arnoldus de Villonova, in the following words: "Whosoever does not understand or know the aforementioned and also the roots of the MINERALIUM, which are compounded by Nature, does not understand as well the basic and natural beginning of this growth." Therefore, it follows that he also knows the Art so much the less. For, as the excellent Magus and philosopher Aristoteles Chymista also says, he shall of necessity not achieve the completion of this endeavor. These fundamental exhortations of these very wise men serve us well and rightly in our intent and undertakings. For he himself speaks of the MATERIA of the blessed Stone, as this is explained in ROSMAJOR, fol. 219: namely, that all error arises because the correct ground and origin of the true Materia Lapidus is concealed and sealed up from them. And it follows from this (he says further) that one who does not know the correct beginning will never achieve his desired end. For he who does not know what he seeks does not know what he will find; and therefore, all who seek without a true ground must doubt and vacillate until Almighty GOD shows them another means, which however happens most rarely. Thus all philosophers truly warn and admonish anyone who wishes to undertake to prepare himself in secret for this excellent and divine Art, that he should first take notice of the Seriem Cursum, or the course of Nature: How out of herself Nature gives birth to and propagates the metallick root, which the Artist must imitate like a monkey.

For the Art can neither make nor create the true thing at the very beginning. No: It is already made by Nature, so he does not need to make it (as shall be demonstrated in what follows). He is not a master, but rather a servant of Nature, for he must serve Nature and come to her assistance. For Nature cannot very rapidly separate the impure from the pure inside the earth, as the Artist can outside it, since the outwardly stinking Sulphur must be sep- arated from the true core, as shall soon be shown. For this reason it is extremely necessary that we should follow the earnest warnings of the philosophers, that is in regard to, how Nature generates and gives birth to the metals in the earth, which every true Artist must follow. For all philosophers imitate nature and everything brings forth its likeness and loves its likeness, and, as Hermes Rex Magnus says, "No extraneous thing, which has neither originated from nor been compounded from the metallic MATERIA, has the power to bring metals into being nor to change or transmute the same." This is also confirmed by Count Bernhard (in fol. 20): "Each substance has above all the seeds particular to it, from which, it comes forth, and it is only propagated through those seeds, and through no others." He continues (in fol. 21): "Such a thing is born and propagated through the metals." But before I pursue this point, with all its prerequisites and other pertinent matters, I think it well that I should first discuss the entire Seriem Generaliter, and thereafter each thing i.n specie, in accordance with the order to which I have committed myself in this work, for the SPECIALIA may be recognized so much the more easily EX GENERALIBUS, and then be taken into consideration.

However in the description of this exalted Art and divine Wisdom, which is properly a secret of natural philosophy, what must above all be considered and observed is the SUBJECTUM, the MATERIA or CHAOS IMPURUM, as it is called by the philosophers, which Nature has produced in the earth in a metallic form, but has left unperfected. This the Artist must gradually purify without heat. For as soon as the outward warmth becomes greater than the warmth within, then the metallic Spirit flows away and cannot be introduced into the dead body, as one can perceive in all metals, which are all dead, since their life has escaped them in fire and has flown away. For this reason they have neither force nor power. The entire first operation or ACTION, up to the (stage of) COMPOSITION or compounding, is nothing more than the sublimation of the MATERIAE, that is, making it subtle. For, as Hermes says: "the crude must be made subtle." In this process occur many operations which the philosophers, for the help of those lacking understanding, place under separate chapters, such as PURIFICATIO, SUBLIMATIO, SOLUTIO, MUNDIFICATIO, SUBTILITATIO, and etc.

This is indeed at bottom nothing other than a purification of the remaining sulphuric odour, followed by a dissolution of the body, so that it may be turned into Sale Metallorum or into Aquam Philosophorum. This, after its external purification, is beautifully pure and virtuous; indeed much more exalted than common gold and silver could ever hope to be (for the spirit of its life remains in it, and this is lacking in common gold and silver). But when this purified Mercury or SAL

METALLORUM is now brought together with its own kind, then the water or Mercury first endeavours to dissolve the earth, so that the earth may enter into the subtility of the water, which may occur when the nature and quality of the water has swallowed up and overcome the earth. After this the earth again begins to have an effect on the water, through a rising and falling action, in order that it might become thick, and in this way to insure that it remains afterwards preserved even in all kinds of fire. This occurs as soon as the quality of the water is overcome for a very great part of our mastery consists in the dissolution of the body or CORPUS in water (although this occurs fully only when the Composito is completed), which the philosophers call a Putrefaction, a rotting or corruption, without which the cyclic transformation of one metal into another may not occur. (ZACH. fol. 78, "quam ob causam" attests to the same thing). For the destruction of the one is the birth of the other; especially when such rotting or corruption and birth have their origin and beginning from a single source. For all metals arose out of one root, as shall be discussed in detail. The Stone of the Wise arises from an insignificant thing, and yet out of this there comes the most noble of treasures; that is, out of the Spermate or seed of our gold, which is sown by means of the Conjunctio or compounding into the mother of the Mercury, there then comes forth the next MATERIA, out of which a higher, more valuable treasure is produced.

This very next MATERIA (which can be recognized when Nature first elevates the proper thing in the earth) is the natural or real moisture, which is introduced into the process of birth by both of its parts (or both of its parents) in their mutual compounding and participation. For only such radical moisture or vapours of the body and of the spirit are the essential parts of the LAPIS or Stone. Thus out of two natures a third cannot be born, unless one of them in the role of the Agentie of active being and the other in the role of the Patientis or guiding being support one another in their efficacy.
From this it now follows incontrovertibly that from the impure things, which have their origin in the well or source of the first minerals, one must exhume, extract and remove the SUBJECTUM or the MATERIAM ELIXIRIS or the highest medicine which will transform or make perfect the imperfect metals, and that one cannot take this out of any other thing in the world, for it can only be made from the mineral MINERA, out of which all metals grow and have their origin. However, that from which all metals have their first origin will be discussed in its proper order IN SPECIALI COMMEMORATIONE of the birth of the

metals, for it is the will and final opinion of all philosophers that we should take solely the inner proper, pure and simple ELEMENTA, and cook and simmer these with gentle moist heat. And they say that if this occurs in any other way then it is of no utility. They also say: Take the most pure, the freshest, clearest, nearest and best thing out of our metallic ores, and exalt it to the tops of the mountains or to the stars of the heavens, and then bring it again down into its roots, and thus shall everything be completed and you have brought about the Rectification of the thing. But how this comes about will be explained and referred to in the SPECIALI EXPLICATIONE, or the explication of the points of Sublimation.

Notice further that he who would seek and perfect the tincture of the SAPIENTUM or the wise to good effect must first recognize the roots of the minerals. Out of this the exalted Work must be perfected, for the recognition of the origin of the body and its qualities is the thing that makes this Work easy. Thus it is that this tincture or medicine made from corporeal things, in which Nature has been brought into a metallic form or nature which is in full accord with and like to Nature herself, can quite properly be ingested without problems and with excellent results. This point shall hereafter be sufficiently explained and presented in the doctrines of the philosophers, and in truth such a tincture can be sought and found in the bodies as well as in the spirits of Nature, since both are found to be of one nature and quality. However the aforementioned tincture from the bodies is more difficult, and that from the spirits easily and more readily, but not more perfectly, preparable from the imperfect metal. For the white path and the red path are both of one root and foundation, after the FERMENT or AGENS has been added to our Mercury. This is attested to by MORIENUS: "Out of one thing, the work of the white as well as the red is completed and brought about, for there is but one Stone and one effect, which can only be very gradually and gently digested through fire and cooking in a single vessel and thus perfected to a red or a white fixed and incombustible Stone." But this must also be done in such a manner that one removes the form or nature of the great elixir from the power of its immediate materials in which it has been placed and concealed by Nature.

It should also be kept in mind that no extraneous water or powder should come into the composition of the Stone in the second operation. ZACH says in fol. 103: "For 'nothing is better to be compared with another thing than that which is most closely related to it and is of its own nature and quality, and if something extraneous should be added to it, then that which was desired will not be produced at the end of the work, but rather everything will take on another form and efficacy." Thus, nothing may have a proper birth, unless it issues from a thing that is of its nature and quality, for Nature cannot be altered nor improved since in her own nature, Nature enjoys her own particular nature. Nature demands or has an appetite for her own nature; she envelopes it and joins herself to it. She rules it correctly; she confronts it; she gives birth to it. She claims it; she conquers it; she keeps it to herself. She renews; she gives increase; she turns white and red; she tinctures and exalts her own particular nature and essence, as is testified by Bernhardus Comes Trevisanus and other philosophers. Furthermore, it is extremely important that the crude, impure, earthly part of the elements, through which they may be burned or grow corrupt, be removed by means of the PREPARATION, or the separation process of the Art, and thus be completely parted from the pure metallic substance, for otherwise the metallic body cannot be resolved well and unlocked, due to the impurity that has accrued to it EX VITIO NATURALI. For the FORMA METALLICA must be removed, and this takes place by means of gentle cooking and digestion, solution and coagulation, in order that

you may turn it into true QUINTA ESSENTIA, the clear
Mercurial water or the crystalline SAL METALLORUM,
which is really nothing but a true pure Sulphur,
which will not burn because it is only the pure
natural warmth, into which natural moisture has been
poured. For the gently prepared fire has the nature
and quality that it unites and mixes the parts that
belong together very well and very densely per
minima: that is, that the waters are mixed and
joined inseparably and most intimately with the
water, so that it destroys and completely consumes
the disagreeable qualities and transmutes and
changes them to the nothingness of an ash. For this
reason, the philosophers desire that one should
bring the external into the internal and the
internal into the external, in order that one might
master the Art; that is, that one should remove the
crude, earthly, combustible, sulphur-like part,
which appears in the external portion of the
materials, by means of the special techniques of the
Artist, and also should bring the most inner; clear
and pure substance, which in the beginning was
planted in the first root of Nature, into the most
external, by cutting away the accidental corruptible
parts. That is, you should bring the internal,
concealed parts to light, and destroy and discard
the external parts, for they are of no use. This is
possible, even easy, for an experienced Artist to do

because the internal parts of a thing stand at all times opposed and contrary to the external parts in regard to their qualities and characteristics, and for such contrary things there is a single process, in which they are opposed to one another in order that they thus may be more easily recognized, and shine forth more clearly. This philosophical Art requires no special method (as many think, that is, that by means of the Art one can make new gold and silver), since Nature herself is wont to give birth to these things in the depths of the earth. Thus the Artist requires nothing more than that he, as a tool through his INSTRUMENTA, should destroy and remove only the FORMAM of the philosophical SOLIS or LUNAE, depending on how he desires to begin his work, by means of our Mercury, And thus he moves the Nature in the SOLUTIONE COMPOSITII, that it might be awakened again through the fire of the Art, and step once again from the dead into life. However it is necessary in the Artificial cooking of the imperfect body that the external mobile warmth be kept in measure with the PROPORTION, so that the power and virtue of the inwardly working warmth, which brings to perfection, should not be altered in the slightest, by neither too much nor too little, which should be determined circumstantially in every case. For if there is too much heat, then the spirit of life flows away, and leaves its dead CORPUS lying behind it; and if there is too little warmth, then it can never be moved to life or growth. For the internal is a pure, fiery, sulphuric, incombustible essence, which, once it has been fixed, may be called the Light of Nature, for it is the radiance and form of all metals, which makes all bodies luminous and perfect. Therefore, if the Artist does not perceive this light, then he may err on many paths before he attains to the truth. For one is not able to see such light, until what has been concealed is brought into this light and the elements are reversed, as Hermes teaches. One is not able to see our spirit, which gives life to all bodies of metal and is also a natural fire, unless it is revealed to one by the Spirit of GOD, or by a living man. And thus all error arises because there is but a single path to this Art, for, as ROSARIUS says, everything that is good or that should become good is prepared and created in only one way, although one man may employ more details and methods than another in the first operation or process, which occurs for the purpose of the COMPOSITION or compounding, yet after the compounding, one must again give it over to Nature, which brings it to the desired end, as GOD has ordained it. The simpler the ARTIFEX or master makes it, the better and more certain it is, for Nature operates and works only purely and simply, and the master must follow her.

But whatever is evil can be prepared in many ways, although not without countless errors, for as soon as a single improper thing is introduced into this Work, many improprieties and errors issue from that one thing, and this occurs because of that extraneous contrary thing, and, in particular, because one is working against Nature. Everyone must guard himself against this, for it is only vanity.

The perfected elixir or the white tincture achieves a likeness to the perfected metals, that is, as the most powerful and effective form of anything, which, if added in solution to the prepared, that is, purged, but imperfect metals, as their most closely related material, achieves a likeness in that it fixes, perfects and tinctures them to an enduring and eternal form, in the greatest and intensest fire. This is the true medicine for men and the perfection of metals, which brings them joy, renews them and transforms them, and, except for GOD, there is no other medicine which expels poverty and all infirmities of the human body, and can keep one in perfect health. At this time few physicians have attained to this knowledge, although many of them think themselves to be upon the right path. With any of our materials which has not been prepared and unlocked (as explained above) it is not easy to work, for the Preparation is the secret of this Art.

Thus it is the Preparation of the efficacious thing through which anything is brought to the point of its movement, perfection and finality; which means that an imperfect metal is brought to the form of a perfect metal, which may occur only through the aforementioned efficacy, movement, light and warmth. For as soon as the warmth declines or is lacking, then the movement or efficacy of the thing ceases, as can be seen in eggs which are abandoned by birds and grow cold. And as every natural or Artificial effect takes, and must take, its own time, whether short or long, through and in which it is brought to its prescribed termination, thus also everything cannot work beyond its own form. Therefore, as soon as the proper form or essence is present, then the movement or the efficacious thing is complete and, taken further will of necessity become a vain thing, which parts effectively from the MATERIA. Besides this, it is also worth noting that one makes the bodies subtle by removing their crudeness and impurity, until they are spiritual, light and pure; but, on the other hand, makes the spirit corporeal and thick, so that they may become enduring and constant, like the bodies. For this reason the ancients say that unless you know how to make the bodies spiritual and the spirits corporeal, you have not found the correct way or process to the venerable Art. This is nothing other than that one

makes the thick thin and light and the light thick
and heavy, as Hermes Rex Magnus says: "crassum fac
subtile et hoc spissum redito", etc. Here it is
important to note that each thin and light thing is
considered more valuable than a crude, thick thing,
because heavy things cannot rise into the heights,
unless they have already been joined to light
things. But, on the other hand neither can the light
things be fixed or held to the ground except through
the power and force of the heavy, crude and thick
things. For, although the body does not act on the
spirit but rather only the spirit acts on the body;
nevertheless, in order for each of them to have an
effect on the other and to be compatible with the
other, it is necessary that the body or earth and
the spirit (that is the fixed and the volatile) be
compounded correctly, according to the PONDERE or
weight of the wise; however, the spirit must
previously have also been purified to the highest
degree by means of Sublimation or Subtilization.
Following this process, that is, after they have
both been resolved and dissolved, just as water is
mixed with water, they are joined indistinguishably,
one with another, and thereafter remain united in
each other, such that no power of fire will have the
ability to separate them again, no matter how great
it might be. The ARTIFEX must be well instructed in
all of this, and thus in the GRADIBUS IGNIS, that

is, in how he should and must rule and form the fire, from the beginning to the end of the entire Work, for otherwise it is easy to make a mistake.

We will come to speak hereafter concerning this at great length, for all the wise Magi and philosophers give us counsel that one should not cast a tincture on an impure metal, nor make any PROJECTION unless, it has previously been well purified, for otherwise it shall suffer great damages, for not only will the tincture be impeded by the SCORIIS and the stinking sulphuric ACCIDENTIIS, but also the tincture will remain at the top, subject generally to great loss. For the content and intent of the entire Art is so directed that one may confer perfection upon the base metals and bring them to completion. Each tincture is engendered through its likeness among the metals (since they all issue forth from the metallic root) which is to be colored and not by any extraneous thing, which has not had its origin in the pure substance of Sulphur and Mercury. For in order to discover and to demonstrate that for all metals their PRIMA MATERIA is wholly identical in its powers and virtues, and thus that it is a very easy path to transform one into the other - and, also that they are distinguishable only through this, at least in regard to their purification and digestion; namely, that one has been much more highly and purely cooked and digested than the other by Nature - it is necessary that the impure metals be more thoroughly and better purified by the Art, so that those which were originally less digested be more thoroughly digested. Thus in this stage of the Preparation, all accidental and chance ingredients, which make the baser metals impure, will be separated from them to the extent that only their pure, incorruptible substance remains, for otherwise, nothing could possibly be transformed and transmuted into the essence of a perfected metal, nor could there be such a metal, since only in this way may the effects of the active thing be completed and fulfilled in the prepared passive thing.

You should note further that only three items (respective loquendo) are necessary to the perfection of the tinctured Stone, in which, if it is correctly prepared, the mastery of this entire Art consists; namely, the Stone of the Sun, which signifies or includes the Red Lion: Red, incombustible Sulphur; and after this the Stone of the Moon, in which the pure and clear, incombustible white Sulphur dominates, as is explained in CLANGOR BUCCINAE in TURBA fol. 484. "In the lunar SUBJECTO there is a white Sulphur." And, finally, the Stone in which our Mercury contains both natures, the white as well as the red. This is the basis of the entire mastery, since our Mercury is the earth, into which one sows, and which brings forth. It is the third Stone, which is the intermediary between the first two, and includes both their natures within itself: "Nam Lapis Mercurii amplectitur utramque naturam", as has been said. You should conceal completely these three metallic and mineral species from the common uncomprehending and unworthy people and let the fools wander on their own false paths, for they are not predestined or foreordained to this knowledge, and it will remain closed to them until they can bring the Sun and the Moon into one body, which cannot, and must not, occur without the Will of Him who lives from Eternity to Eternity. For this high Art is the highest earthly gift of Almighty GOD alone, and shall be held in His hand, and neither given to nor taken by anyone, except only by him whom He will. Nevertheless, at times it may also be attained through keen and exalted understanding, through prayer, through continued and diligent reading of books and industrious seeking, or through the revelations of a faithful master. Thus it may also come through me, by the grace of GOD, beyond any doubt, for a true philosopher has GOD before his eyes, and does what is proper, or otherwise he had better not do anything at all. Thus I do not make these disclosures for my sake, but for the sake of that which shall be revealed. I thank GOD for the abundant support of his miracles. And indeed, he who has correctly read and understood the CODICES or books of the ancients will not be able to deny that I have not corrupted the truth in this, my methodological theory, and that I have not invented the PRINCIPIA or source and mystery of this sacred Art as will be shown. The eternal GOD gives it only to those who intend to employ it for the good of their neighbor and the edification of the Christian church, in honour to His name, Amen.

Here follow the main topics, in which the true foundation is truthfully brought to light, so that one might comprehend it. And it is easy, and not difficult as several have already claimed. Up to now, following order and necessity, the high ARCANUM GENERALI MODO has been discussed, thus there now follows the basic instruction and demonstration IN SPECIE and in particulars, for which it is necessary first to disclose points relating to the GENERATURAM or coming to birth of the minerals and metals, and subsequently to demonstrate the true materials or SUBJECTI LAPIDIS, with corresponding details, requisites and other pertaining matters: to describe them, prove them and present them. (For the sake of excellence this should and must be done in accord with the course of Nature, that is, according to how the metals and other similar things have their

origin in it, which every true philosopher and
Artist must follow, since all philosophers imitate
her, and write in accord with her.)

However, before I begin to explain this, it must
needs be brought to notice that the philosophers
mention three sorts of minerals. The greater
minerals are metals which still remain in their
CHAOS, husks, or (as Theophrastus says) still lie in
the workshop of Nature, that is, before they are
smelted to metals through the power of fire.

The middle minerals are the Marcasites, and all such
species in which a metallic radiance may be seen,
such as antimony, WISSNUT, arseno-pyrate, etc. No
metal comes from any of these, however long they lay
in the earth, for they do not possess a complete
Blossom or Button, but rather have only two
PRINCIPIA, namely, Sulphur and Mercury. They lack
Salt.

The third sort of minerals are called the lesser
minerals, and are the salts and all such things as
alum, vitriol, saltpeter and all kinds of ground
mass, since no metallic form or radiance can be
perceived in them. It has been necessary for me to
explain this, in order that no misunderstanding of
the term "mineral" might arise.

Now follows the necessary explication of the Generation, or coming to birth, of the metals, which the ARTIFEX must follow correctly, for everything brings forth its likeness, as is to be observed in all living things.

However in this regard you should especially understand that all metals spring from and have their origin in one root, material, foundation and ground, for otherwise they would not be HOMOGENEA or CONSANGUINEA, that is, close blood—relatives. This is confirmed by all true philosophers and is also given by experience; for there is no tin which has not first been lead, all silver was previously lead, and similarly all gold was previously silver, as is attested by the highly enlightened Count von Tervis in fol. 31 and 32. Then he artfully describes how the metals grow, and says that they are first lead, then tin, then silver, then copper and finally iron and gold. However, Nature takes a longer time with iron and copper due to the impurity that they have assumed at their birth. Everyone can read this description in that book, and everything that it reports is as it says. DION ZACH. in fol. 92 reports (as I myself can also attest) that a mine is often excavated before its proper time, in which case one then finds premature silver, whose character is like that of lead ore; but if the mine is closed again, and the ore is allowed to be digested for a certain time (approximately forty or fifty years) then it contains almost pure silver. Although this may appear miraculous, because lead is impure, but silver is a beautiful, delicate, pure metal, it is to be explained in no other way than that Nature casts the impurity out of the lead in relation to the duration of time, as we shall prove. And the inner, Mercury, Salt and Sulphur of Saturn within it are always as beautiful, pure, lovely and good as is ever found in the Moon or the Sun, as shall also be demonstrated.

That all metals have come forth from one root has been written by the true philosophers who have attained the Stone. One must lend credence to an artist or a craftsman concerning what he says he has learned in his craft, not to mention such exalted Magi and philosophers as Rex Geber and CLANGOR BUCC. in TURBA, Fol 473, who speak about it most usefully, as follows: "Secundum varietatem sulphuris et ipsius multiplicationem diversa metalla procreantur in terra." That is, the various metals are born in the earth according to the variety and quantity of the sulphur and also to its propagation. But the first material out of which they stem is one and the same, for the metals are not different, except through the

43

accidental effect of having received a greater or lesser, tempered or untempered warmth and a combustible or an incombustible Sulphur in the bowels of the earth. In this (he says) the greatest number of philosophers agree. But how it comes about will be discussed in the section concerning how and from what the metals are compounded by Nature. TURBA PHILOS. fol. 579, thus says: "The philosophers have had such exalted thoughts, that they have attempted to bring the lower bodies of the planets into correspondence with those that are above in the firmament in terms of their external radiance, light and purity, and they have succeeded in this, because they were established on the ground of truth, since the metallic bodies differ only due to a greater or lesser cooking, while their origin and beginning is Mercury." And thus TURBA, fol. 610, says: "Our Mercury is all metals." In this regard the philosophers always use the plural form, i.e., "metalla, metallorum, metallis." Even when they should say, e. g. "from a metal" or "the metal", they say instead, "the metals", "from the metals", "of the metals", and mean thereby that the metals are very closely related. Count Bernhardus relates this in his parable, when he says that the other six metals come from the well, that is, from Mercury, but they have not yet attained exaltation, as he has. It would take too long to quote all the

44

authorities and proofs in both languages, and thus I will only cite for you the relevant passages, so that you may read them yourselves. They are: TREVIS COMES; fol. MIHI 44; FLAMEL: fol 119 in fine; ARNOL. in ROS. fol 399, fol. 411; MAGISTER DEGENHARDUS in SUO TRACTATU DE LAPIDE, fol. 116; HOLANDUS in LIBRO VEGETAB. in PUCTO SATURNI, fol. 212; BERN. AGN., fol. 29; TURBA 277; CLANGOR BUCC. fol. 437; B. AG., fol. 109; TURBA, fol. 177; and similar passages. And in truth and in fact it is not otherwise.

Now we may properly discuss from what and by means of what Nature generates or gives birth to the metals in the depths of the earth. In this regard you should understand that Nature has veins and passages in the earth, where salty, clear and unclear waters drink, lick, drip and deposit, as can be observed in mines, where acidic, salty water always drips. That is, whenever the salty waters press down, from above, but it is not true that heavy things always sink into the depths, because the sulphuric vapours meet them from below out of the center of the earth. Should it then occur that the salty waters are pure and clear and that the sulphuric vapours are also pure, and that they meet and join with one another, then a good metal results. But when this is not so, then an impure metal results, on which Nature has to work for more or less one thousand years before she makes it perfect, and this is due to the impurity of either the Mercurial salty water or the Sulphuric impure vapor. If the two have received one another in a closed cleft or stone, then there rises from their junction a moist, dense, fatty vapour. Because of the true warmth of Nature, this condenses, since it has no air (for otherwise it would flow away), and from this vapour there comes a mucilaginous or greasy material, white, like butter, which Matthesius calls a GUR. This may be spread like butter, as I can also clearly demonstrate here, on or above the earth. This GUR is also often found by miners, but nothing can be made from it, for no one knows what is available to Nature for such Work, and it could as easily become a marcasite as a metal. This compounded material is thereafter brought to a metallic form, or MASSAM, by means of the long, gentle and vapourous cooking of Nature. The first form of the metals is a lead-like material, in which a kernel of FIXI of the Sun or of the Moon is always concealed, which acting as a seed, always grows and hastens toward the perfection of the Moon, for which reason it is properly called LUNARIA, but also SONNENWERT. Therefore, Flamellus says in fol. 118, that one can observe in leadmines that it is impossible to find lead from which one is unable to extract at least a grain or kernel of gold or silver. Count Trevisanus confirms this sufficiently in fol. 31 and 32, where he establishes the order of the GENERATIONIS METALLORUM, with lead as the first, and then tin; but all the books of the philosophers are full of such descriptions. However, due to my previous intention, it must be proven that the generation of the metals proceeds as I have said. CLANGOR BUCC. speaks also of this in fol. 473. "For every one of these metals is compounded first from Mercury and Sulphur, and then transformed into an earthly substance, after which a subtle, light and pure vapour rises from these two earthly substances." Pure metals come from this vapour because of the internal warmth which is found in the cavities of the earth, by means of which it is digested and cooked, until it is all transformed to an earthly substance and nature. It finally becomes fixed (after it has lain long in the workshop) and is transformed and converted into a metallic nature.

NIC. FLAMMEL; an excellent philosopher, writes usefully about this in fol. 152: "It is certain that no extraneous or imperfect thing makes the imperfect metals perfect or transmutes them." For this reason, people can properly be held to be irrational who think to produce something in this Art, out of animals or any green things, since they may use minerals, which are very closely related to the metals. For only from these two, namely Sulphur and Mercury, are all metals born. But in this connection, let no one make the mistake that we are speaking only of two things, and thereby leave the Salt out. For you must understand that the Salt is concealed in the Mercurial water, and that the water can quickly and easily be transformed to SAL METALLORUM, and that the SALT can again be turned to water. In addition, it has also been related that the metals are engendered from a salty vitriolic water and Sulphuric vapour ("semita semitae," in TURBA, fol. 473). You should understand and note that the Mercury is cooked in the earth, as follows: the son of all metals is imperfectly digested in the bowels of the earth by means of Sulphuric warmth or vapours, and, in accord with the varying quality of this Sulphur, that is, according to how pure or impure the Sulphuric vapour is, the various metals are born in the earth. But their original MATERIA is of one kind only, and they are different only in so far as one is more cooked, or is more burnt by impure Sulphuric vapour than the other. Thus they are variously engendered, and in this all philosophers agree. It would take too long to quote here the proofs of all the philosophers verbatim, especially since they all agree with one another;

you may read Count Bernhardus in the cited place yourself. The passages are:

TURBA, fol. 495, 356, 476; FLAMEL, fol. 183; CLANG. BUCC. fol. 493; TURBA, fol. 411; VERB IDEOQ, fol 569 in fine, fol. 31, 32, 40, 44; MAGISTRU DEGENHAR. fol. 122; RICH. ANGLIC. fol. 127, 310, 579; ROSINUM, fol. 278; ARN. in FLOR. fol. 475 in fine; TURBA. fol. 158, 159, 160; FLAM. fol. 152. They are all concealed in the SALE METALLORUM: body, spirit, soul, Sulphur and Mercury, and in order to demonstrate this, Hermes says: "Sal Metallorum est Lapis Philosophorum et qui habet sal metallorum ille habet secretum sapientum antiquorum." All of these men and all of their beliefs give one voice to this: That all metals are born from Sulphur and Mercury IN QUO SAL LATET; and such is indeed the case.

Because I have now sufficiently explained the place and quality of the MODUM, and have demonstrated that from which the metals are generated and born, and because every Artist should, and must, follow this, thus it will now be necessary to show, speak about and describe the correct and true MATERIAM, so that he will be able to follow Nature, since this is of the greatest importance. For it is useless to build if I have no material with which I intend to build. However, someone at this point may make an objection and say: Although you can name or show me a MATERIAM

49

without any trouble, who is to know whether it is
possible that the metals grow and propagate themselves or can be propagated and multiplied? Because
this is a very important point, we must now leave
the explication or nominal interpretation of the
materials, somewhat behind us, and present, explain
and demonstrate this point with fundamental and
unobjectionable arguments and reasons.

That very wise and intelligent man, Arnoldus de
Villanova in LIB FLOR. FLOR., provides the first
incontrovertible argument and proof in these words:
"Everything that grows also propagates itself, as
can be perceived in trees, grain and all things. But
the metals grow; therefore they also, like other
things, can be multiplied and propagated." That they
grow can be observed through experiences such as
those mentioned above: That a mine, in which the ore
has been found to be still imperfect, can be closed
and allowed to grow to perfection through Nature, as
all experienced miners know. Almighty GOD Himself
speaks for this, since there is increase in the
creation of all things, for he says in Genesis I:
Let everything bring forth fruit after its kind, and
be fruitful and multiply itself; thus, there can be
no doubt about it. An excellent philosopher who possessed the Stone, namely Dionysus, speaks as follows
in fol. 78: "Everything which is ordained for

completion and perfection, and yet has remained imperfect due to a lack of cooking, may be brought to perfection through subsequent cooking."

The perfect metals are predestined and foreordained to perfection; for this reason they can be brought to perfection by means of a steady subsequent cooking or digestion. This argument is directed principally towards our MATERIAM, for this is also left imperfect by Nature, and the Artist must thus come to her assistance with purification and cooking, as shall be explained. The following third argument is from the certainty of TRANSMUTATION or change, or, as one now says, of tincturing, and is presented by ARIST., 1.4 in METEOR. and ZACH. fol. 79, as well as COMES BERNHARD., ALB. MAGNUS and Avicenna. It goes: The alchemists cannot transform any metal into another unless they reduce it first into their first MATERIAM. This reduction or regeneration into the first MATERIAM, is possible and very easy, as a result of which the transmutation or transformation is also possible and easy. I should possibly relate much more concerning this reduction or reproduction into the first MATERIAM; however, since my Edifyer, that is, this tractatus written by me, has said enough about it, and, in addition, it can be shown that the books of the philosophers are full of references to it, (such

as Count Bernhard in fol. 17, 18 and 19), I think
that I have already satisfied you in this regard.
This point is to be well noted, for many people have
erred therein and think that when they have the
MERCURIUM PHILOSOPHORUM or the SAL METALLORUM, then
they have also the PRIMAM MATERIAM. No; but rather
it will first become the PRIMA MATERIA after the
Composition has occurred, of both the Man and the
Wife, as is attested by the Count in fol. 21 in fine
and 22. There he says: "In principio turn demum ista
conjunctio dicitur prima materia et non prius". Then
only after the conjunction or composition is it
called the first material of the Stone, or of all
metals. In this connection read TURBA, fol. 415, 364
and so on.

Thus, legitimately and on the basis of Nature, that
is, from rational causes, we cast out and reject the
ANIMALIA and VEGETABILIA as extraneous, contrary and
unsuitable materials for our Work, and establish our
MATERIAM LAPIDIS properly and correctly among the
MINERALIA, since all metals are smelted in minerals
or from the same (unless they are found in a pure
state, in which case they have released themselves
from the mineral, as chicken from the eggshell) or
are extracted and purified by other means. But now
we are confronted by the following question: Since
we have already established the tripart division of

the most excellent minerals, that is, the greater, middle and lesser minerals, out of which of these, then, should our MATERIA principally and most successfully be taken, or, to phrase it better, from which of our minerals or metallic ores does the proper MATERIA come? That it must be metallic we shall prove in what follows; for everything brings forth its own kind, takes joy in its own kind and hates its contrary. The philosophers produce much confusion in their description or nomination of the true materials. For instance, FLAMEL says (fol. 152 in fine) that it is excellently concealed from which mineral MATERIA our Stone can be made, but that in part it is very closed, in part very insufficient in material, and also that most of it is impure, for which reason many fail precisely at this point. CLANGOR says (in fol. 475) that he obtained only a quarter ounce of Mercury out of a full pound which proved suitable for his work. Arnold de Villanova also writes of this in ROS fol. 404, as well as ZACH, fol. 433, 92, and 150. For there are various minerals, among which some are more purified, cooked and digested by Nature, and these are closer and more suitable for our Work, and also better, as I hope, my dear friend, that you will know. For what Nature has already made, I may not make, and this is all to my advantage.

But in addition you should also understand
fundamentally that the MATERIA LAPIDIS or the LAPIS
PHILOSOPHORUM may be prepared from all the metals,
especially when they are still contained in their
minerals; but that as soon as they come into fire,
then the SPIRITUS TINGENS flows away, and leaves its
CORPUS lying dead behind it, as is related in
ROSAR., fol 209. Our Stone is a thing or MATERIA
which never comes into a fire. With the SPIRITU
CORPORALI that flows out of the metal I can tincture
a Venus in an instant to the appearance of the Sun,
but this is not enduring, because it is a fleeting
spirit. But if the material is raw, then there is a
possibility to do whatever one desires, if it is
executed highly and nobly through the Art.
Nevertheless, there is a shortcoming, for some
metals are bound too hard, and some are too impure
and difficult to dissolve. For this reason, the
philosophers have chosen the nearest and easiest way
and have taken the materials that can most easily be
unlocked, and also those in which the PRIMUM ENS and
VIS GENERATIVA ET MULTIPLICATIVA are still
plentifully contained, in order that they may
achieve their end all the more quickly. Especially
since the MATERIA is the same in all metals, why
should I plague myself for one or two years with a
Resolution which I can otherwise produce in the same
materials in at most four to six weeks, since I can

have the same MATERIAM in a short period or a long
one? That this is true is confirmed by the very
famous philosopher, Avicenna, in fol. 433, with
these words: "One should know that some metals can
be made into an elixir much more easily than
others." TURBA (fol. 404 in fine): "I say that all
metals are internally gold and silver, which
everyone who understands this Art is able to know."
FLAMELLUS (in fol. 120) says even more clearly that
the Stone can be made from all metals (as long as
they have not been in a fire). You can find it also,
he says, in LUNA, or where ever you desire to seek
it, in lead, iron, more certainly in copper. But I,
he says, have found it in gold (but understand that
he does not mean common gold). In all of this it
must be understood that it has not yet been in any
fire. This distinction is made by Count von Tervis
(fol. 16 in fine): "Leave all the minerals alone
(understand; the lesser and middle minerals) and use
only the metals. However, most of the Artists do not
understand this passage, and this makes a great difference, for when the metals go through fire and are
smelted into a metallic form or MASSAM, then they
are single and alone, for they have lost their
SPIRITUM TINGENTUM and are now dead." In addition,
it is in this way that one body is separated and
driven from another; for one can only rarely find
one metal alone, but rather Venus and the Moon are

55

gladly together, the Moon and Saturn are gladly
together, the Moon and the Sun are gladly together,
and so on. Thus they plague these compound bodies
further and separate first Venus and the Moon,
Saturn from the Noon, the Moon from the Sun, that
the metals thus become powerless, impotent, dead
single bodies and can do nothing. Even the Sun can
do no more than what it possesses. Thus they become
separate, alone and no longer a compound of body,
soul and spirit, for as soon as the spirit leaves
its body, then the soul also becomes powerless, and
it is therefore impossible to make anything from
them. Let us approach this more closely. Previously
we discussed the three sorts of minerals, the
greater, lesser and middle ones. The lesser ones
cannot be used, for they are only marcasites and
contain only two PRINCIPIA, as we have stated
previously. Thus the Count has rejected all of the
minerals that do not bear metals; indeed, he rejects
even the metals themselves when they have been
sundered from their life.

At this point someone may ask; Since there are so
many minerals which contain metals, how do I know
which are the most suitable and best, from which I
can easily obtain our Materium or the Mercurius
Philosophorum.

The philosophers themselves address this point or question when they say that we should use that mineral or MATERIAM which Nature has already begun to make into a metal, and thus has brought into a metallic form or radiance, but has left uncompleted; we should take such a material. Now, concerning the generation of the metals, we have previously heard that Nature first makes a mineral lead ore, as the Count relates: the first is Saturn, and the second is Jupiter. However, one may not be willing to trust me in this regard, and thus I will support my opinion and explain the true MATERIAM, which is most suitable, so that one may have no doubt about it at all. Therefore, extend your reason, open the eyes of your understanding, and then pray that you may be able to comprehend it. For it is my experience with the MATERIA that you are to learn in this edifying tractatus, although everyone may not be able readily to understand it. However, I will endeavor, as is proper, to make the clear truth as clear as I can, and may GOD make you silent, that you may not reveal it capriciously to another man or to him who is not worthy of it. AMEN.

I have previously mentioned that I wish to prove that our blessed Stone arises only from the metallic root, and thus must be a metallic body, so that it may rectify, cure and tincture the impure metals

57

into its likeness. This may serve to insure that you do not seek it in vegetable, or animal substances, even though our Stone is also animal and vegetable, for when it is resolved in water, then it is called "aquam vegetabilem, nam vegetat proprium corpus," for it brings its own body to the process of growth. It is animal because it has an Animam, as is explained by ZACH, and others, and this is its life, which flows away from the common metals in fire, so that they become dead, since their spiritual essence or tincture is gone, which should have tinctured them. For a body cannot penetrate or tincture another body; the spirit is the driver; it must do it.

That the Stone must be metallic is proven by ROSAR. MAGNUS in fol. 2;31, where he says that the MERCURIUS PHILOSOPHORUM is that on which Nature has worked but little, or has brought it to a metallic form or radiance, but has left it imperfect. Also ROSAR. fol 252: "Our Mercury is not just anything, but rather that upon which Nature has completed her first operation and to which she has given a metallic nature, but has left imperfect." Also fol. 304: "This alone, because it is metallic, contains in itself that which is necessary to our work." CLANGOR BUCC. fol. 476: "Epimidius philosophus in turbae per colotem temperatum extrahitur a materia."

That is, by means of a gentle warmth, a slimy or vaporous mucous like is extracted from a metallic material. But it must also be subtle — "hoc est sublimatio philosophorum, sicut Hermes dicit, crassum fac subtile et hoc spissum redito." They call a mixed and well purified earthly substance an elixir (which is a general medicine for men, animals, trees, metals and plants) by means of which one can transform the metals. It would take too long to quote all the authorities and proofs in this place, so I will only cite the references, and you can refer to them and read them yourselves: ARNOLD. in ROSAR. fol. 405; FLAMELLUS, fol. 137; CLANGOR, fol. 475, 510; RECIPIAMUS; FLAMEL, fol. 1419 Lapis; TURBA, fol. 155, 433; but also note: TREVISANUS COMES. fol. 21, 35. These should suffice so that one may not doubt that the MATERIA LAPIDIS is or has a metallic form and character, which must, however, be taken from it by means of a Solution, as will later be discussed. Because it is a metallic body it cannot tincture; therefore, it must be brought into a spiritual form through SUBLIMATIONEM PHYSICAM, in order that it can penetrate and tincture.

There now follows the CONCORDIANTIAE or general agreement of all philosophers by means of which, I will prove that that which I shall name hereafter is the most suitable material for our Stone. I will cite various texts, some of which are more relevant than others, but in the first place I will introduce the text through which the eyes of my understanding were opened through the mercy of GOD, and this is found in FLAMELLO, fol. 118. "When the Mercury of the metals congeals and stands only a little while in the first material, then there is soon present within it an enduring Kernel of gold, which brings forth and procreates a true sprout of our Mercury from the two seeds (SULPHURE PINGUI ET SALE), as can be observed in lead mines, for one can find no lead ore out of which an enduring Kernel of gold or silver cannot be brought to light." For the first binding together or freezing together of the Mercury is a mineral of Saturn or a lead ore, in which it has been introduced by Nature. This is also confirmed by Count Bernhard in fol. 31. This can truthfully be brought to its perfection or completion without any doubt or error, but this fixed Kernel must remain contained in its Mercury, and not be separated by anything from its mineral or lead ore, for otherwise it will be extracted by the power of fire, turned into silver, and become useless, as follows. For when a metal still lies or is contained in its mineral or workshop, then it is a Mercury, which, if its fixed kernel is separated from it, is like an unripe apple which is plucked from its tree and then completely rots. For the fixed kernel is like the apple and the Mercury is like the tree, and one must not part or separate the fruit from the tree, for it can have no other sustinence than from its Mercury. This text is indeed so clear and correct that even a very simple man can understand it, and this text was my first enlightenment, which occurred to me through the grace of GOD. And from this one passage alone may the true foundation be received; for all the philosophers write that one should take a material on which Nature has first begun her work and brought into a metallic form, but has left uncompleted. This he also says in this passage, for the first form of the metals is PLUMBAGO, as is also confirmed by Count Trevisanus, when he says that the first material is lead, for this has already been in the fire, and has lost its spirit or power of life. This spirit can also tincture copper into an appearance of gold, which, however, is not enduring. This means that when the entire body has already begun to be developed by Nature into a metal, then the possibility exists that the Spirit can work with the raw material according to its will, if it can first be brought to exaltation through our Art. And this SPIRITUS must also have a body, which is the LUNA or LUNARIA and which is hidden in it, but can be revealed in the SALE METALLORUM. Thus Hermes says: "Its father is the Red Sun, but its mother is the white Moon." The white LUNA, he says, in distinction to the common LUNA. For our Moon is transparent and yet the LUNA is contained in it, which can be proved, for the FORMA METALLICA must be removed from our Moon, or otherwise it could not mix PER MINIMA, that is, be mixed thoroughly, like water with water. There is much I could say about this, but I must endeavour to be brief. However, the resolution of the body or material, when the Salt is prepared, must be repeated again and again.

Now other texts follow. FLAMELLUS, fol. 126: "In the earth a lead-like material, MERCURIUS COAGULATUS, grows. This one should cast into prison, and then execute, for in this way one can discover its weight, which is otherwise difficult." This lead—like material should be enclosed in a vessel and then purified, and only thus can one determine its weight. This is his opinion, because when it is still impure it is impossible to assess. Thus, for the facilitating of our Work, there is only one metal to be found in the entire world, from which our Mercury may be plentifully obtained. Saturn, that is, lead, is heavy and soft, for which reason it is comparable to gold, and is called leprous gold. For indeed, he says that lead which has not been in fire is leprous gold; its leprosy, is removed from it in solution, so that it is made clear, as the substance of gold must always be. Thus CLANG. says in TURBA, fol. 502: "Our ore has a leprous and dropsical body, like the Syrian Naaman (2 Kings 5). Therefore it desires to bathe itself seven times in the Jordan, in order that it may be cleansed of its congenital impurity." FLAMEL., fol. 116: "Therefore climb up the mountain, that you may see a vegetable Saturnian, lead-like, royal, mineral root or plant; take only the sap of this and throw the husk away." He cannot evade it; he must call this lead MASSAM a royal MATERIAM, for in it and from it, gold is secured and born, as has already sufficiently been stated. We want to hear as well what that excellent man, THEOPH. PARAC. thinks of this and what his MATERIA is, for he says in Libro Vexation fol. 38: "For thus speaks Saturn of his own particular nature: they have singled me out from the six (understand the metals) for their experiments and driven me out of the spiritual city; they have locked me out of our dwelling with a corruptible body, for whatever they do not want to be or to have, that I must be. My six brothers are spiritual, and thus they go through my body whenever I am fiery, and I vanish in the fire (understand here the crucible or other experiments) and thus they vanish also with me. Except for two of the best, gold and silver which they cleanse to beauty by means of my water and then become proud. My spirit is the water, which softens all the frozen and stiff bodies of my brothers, but my body is drawn toward the earth, so that whatever I touch becomes like the earth and is made by us into a body. It would not be good for the world to know or believe what is in me and what I am able to do. It would be much better if they would do with me what is possible through me and leave all the other arts of Alchemy alone, using only what is in me and what can be accomplished with me. The Stone of Coldness is in me: that is my water, with which I may fix and freeze the six metals to the corporeal essence of the seventh; that is, to promote gold with silver." This text is clear enough, and stands in need of no interpretation, except only that it is not to be understood as referring to common lead, for this has been in fire.

Here follows a text which I have copied from THEOPA in my own hand. He says: "Therefore I tell you that you should take the sickest of the seven sick ones (that is, Saturn). For him it is needful that you lead him to his bath of purification and purify him of that which Nature has planted in him against his will. Thus you have a secret...." etc.

Arnold de Villanova in L. NOVA LUM., fol. 457, says: "I testify to you that such possibilities are found

in the MATERIA, from which, with my own hands,
witnessed by my own eyes, but through the doctrine
of another, I have made the elixir that turns lead
to gold. This I have named for you, and it is the
Philosophorum Magnesia, from which the philosophers
have extracted the gold that was concealed in its
body." This stands in full agreement with the
previous citations.

MAGISTER DEGENH., a monk of the Augustinian Order,
who certainly possessed the Stone, says in his book,
DE VIA UNIVERSALI: "It is a thing which is like all
things internally, and in it all earthly
(understand: sublunary) things are also concealed.
Its tincture against all diseases is a gift of the
Holy Ghost. In it lies the secret whereby one may
attain to the Treasure of the Wise, and that is the
Plumbum Philosophorum, otherwise called the Plumbum
Aerie, namely, lead ore, in which lies concealed a
lovely, radiant white dove, called SAL METALLORUM,
in which the highest mastery of the Work consists.
This is the chaste, intelligent and rich Queen of
Sheba, clothed in white VLIAND, who will subject
herself to no one but the wise King Soloman. No
human heart is able to fathom all of this." At the
end of his tractatus he also adds: "It is truly a
wise man who can recognize the nature of lead." He
speaks very well when he says that the white dove

hidden in it, lies within metals which have not yet been in any fire. But the philosophers agree almost unanimously that it lies in Saturn, because the body of Saturn is the easiest to open and unlock, as has been previously mentioned. Therefore, as we have said, one must take the kernel and throw away the husk, as the DIALOGUS PHILOSOPHIAE so well explains in fol. 14 and 16: "One must not take that out of which the metals have come. No: But rather that which has been extracted from the metals." And this is precisely that which lies concealed in the metals, which is explained so excellently in TURBA, part 1, fol. 577, where it gives the example of a tree: "If one wants to grow a tree, then one must not take the water or the earth but rather that which is of the tree, namely, the seed or sprout, which one again gives over to Nature." You can read further concerning this in RIPLEUS in AXIOMAT., fol. 179: "The compounded metallic clump is compact galena (BLEISCHWEIFF), for which reason we call it lead (Blei); the quality of its radiance comes from the Sun and the Moon."

Hermes the great King and Father of the Philosophers, says in his book, DE CHAO GENERALI (chapter 19, fol. 268, n. 14): "The most excellent purification of our Mercury is that one removes from it its leaden darkness or form with the help of wine, so that it is made glorious, clear and translucent, like crystalline transparent Salt. This cannot be done, unless the FORMA METALLICA is removed, so that a spiritual essence is created, as is intended in the solution." DIONYS. ZACH. says in fol. 92 that we should and must take precisely the same material from which Nature makes metals in the earth. Count Bernhard explains in this regard that Nature first creates a lead-like MATERIAM in which she establishes her first DISPOSITION and quality. From here one can pass to the next point, which is that the other bodies are too hard to be unlocked. FLAMEL also testifies to this in fol. 153 and 154, where he says: "Why do we not take the pure bodies of the Sun and the Moon for our Work? The answer: because Nature has bound them together too hard, so that one cannot easily obtain anything from them with fire. Instead we take a body which contains precisely as pure a Sulphur and Mercury as are in silver and gold, on which Nature has only worked a little, but has left uncompleated." He states further that we should take only that which is not completed, for the Sun and Moon are perfect, and therefore they are already in their final state. One may also read where he circumstantially explains the MODUM MATERIARUM; and it is likewise worth noting that the philosophers always use the plural form, as in "bodies" METALLA, METALLIS, METALLORUM" and do not say "a body", "a metal", etc. This occurs only because they want to lead the unwise astray, for all metals, as we have shown above, come forth from one root. Thus it is, in particular, that the Universal can be prepared from them before they have been in fire, and especially if they are still contained in their mineral. Therefore, ROGER BACCHO and FLAM. fol. 137 say: "Nothing may be joined to the metals, neither may they be compounded or transmuted, except only with that which issues from them", as we have previously explained.

It would take too long to explicate all the texts; we wish therefore simply only to carry out the proof; for there can be no doubt concerning the MATERIA, as will be shown in deed and in practice. RASIS says: "all secrets are contained in lead", but not in common lead, for he adds: "You should not understand this simply as common lead, in order not to err, but rather as our fragile and black Silberglett. And truly in our lead then is potential gold and silver, and not only the visible whiteness." This passage is clear enough: that the Moon is contained in it can be demonstrated in practice, even though one cannot see it when it is compounded in it. For when its spirit departs from its body, it leaves the lunar body behind it like the loveliest silver, in every sample. Now it is undeniable that all gold was once silver, and that gold can easily be made from silver. Thus LUNA or the LUNARIA can be shown to be here, and for this reason such a process can easily be brought to PLUSQUAMPERFECTION through the exalted Operation. MARIA PROPHETISSA, the sister of Moses, says in TURBA, fol 322: "The fixed or enduring body is from the Heart of Lead", that is, from the internal substance that is contained in our lead. As previously stated, a grain or so of the Sun or the Moon is always contained in Saturn which has not yet been in fire, like a seed that may be propagated. Thus ROSARIUS, fol. 265, says: "Our Sun or Moon, or the fixed Body, is concealed, just like the soul in the human body, or like fire in wood or in stone." AURORA CONSURGENS in TURBA, fol. 220: "Behold, I have shown you the composition of our white lead (that is, when it is turned to a white salt by means of Resolution) and when you know this, then the other is woman's work and child's play." He will thus be understood to mean that after the Composition it is the easiest work that might be. Count Bernhard also confirms this in fol. 3: "Our work is so trifling and simple, indeed so easy, that if I should tell you about it with words, or show you with deeds, then you would not believe it." ZACH. also says: "If the philosophers had observed the correct order, then one would be able to

69

understnad this Art in a day or an hour, since it is
so noble and simple." However, the wise man should
nonetheless consider how Almighty GOD has placed
such an exalted Work in such an insignificant,
contemptible SUBJECTUM or MATERI (for He has at all
times a desire for smallness, but also so that the
rich may not take notice of it, although they have
nothing but gold in their hearts, for they cannot
recognize or perceive this insignificant gold) for
then how much more can He bring about with more
exalted things!

TURBA, Part 1, fol. 221: "In lead is a living death.
And what the philosopher says should be included
among the secrets of all secrets: nothing is closer
to gold than lead." This can be more clearly stated:
the concealed gold lies dead in the lead, but when
its death, that is, its impurity, is taken from it
by the Solution, than it becomes alive and finds joy
in its own kind, to which it is then added. For it
is like wax, into which all seals may be pressed:
if added to gold, then one obtains gold; if added to
copper, then it brings forth copper, and so forth.
This is confirmed in ROS, fol. 319; TURBA, fol. 406;
ARNOLD in ROS., fol. 411; and TURBA, fol. 59: "Own
sole sit Sol, cum Luna Luna, cum Venere Venus." That
it can accept all forms into itself is confirmed by
FLAM. fol. 168, who states: "Mercurius, induit

onines formas, sicut cera omnia sigilla." TURBA, fol. 39: "In our earth are three eyes; the rising, the setting, and out of them our white Saturn is born, which is the SAL METALLORUM." Also TURBA part 1: "Our white camel is the seventh in the number. According to the exalted philosophers, the Sun and the Moon, as well as Jupiter, our Mars and Venus, are contained and gathered in our Mercury, but Saturn is the seventh, in which all of them are contained and united. He is the spatula, the sword, the knife and the incision of that which is born in miracle, with which one can resist one's enemies, and in addition a cask of good wine." What could be more clearly or distinctly said; he even provides additionally the agent of dissolution, for this is contained in the wine cask. This is, in the first place, wine, which is Spiritus Vini; but it is also Acetum Vini, Sal Tartari, and other things as well, in addition to which other agents may be used. These, however, may not remain, but must be removed again after the Solution.

ARNOLDUS in FLOR. FLOR., fol. 471 in fin: "Metalla non generantur nisi ex spermate proprio", the metals may not be generated except from their own seeds. This has already been sufficiently discussed, only one must perceive what the first metal is. For this read TREVIS, fol. 31 and 32 and TURBA, fol. 389 in prin. part. Our old man who seems to be dead is a figure of our science (understand in him the old Saturn). In him the composition of the natures (that is, earth, water, fire and air) is perfect; and all of these are in Saturn. Through him the Doors of Wisdom, the Doors of the Seven Metals are unlocked and opened, as Hermes and his ancestors teach. What could be more clearly said than ROS. fol. 394: "Praised be the Creator of all things. Who, out of something insignificant, has created something precious, worthy and exalted; and, since this is metallic, it contains within itself everything that belongs to the entire work." Here he says that it is taken out of what is most insignificant, which is Saturn, in which Nature has planted the first metallic form, which has been sufficiently discussed already. ARNOLDUS in LIBRO DE CHAO UNIVERSALI ASTIPULATOR in AURORA CONSURGENTE in TURBA, fol. 203: "The seeds of our science are extracted from a metallic body, which contains most potently within itslef the power and virtue of all metals." You have heard enough already concerning what sort of a material this is, in which all metals are contained, so I need say no more. This text is as clear as brightest day.

Magister Degenhardus, Lullius and Matthesius, in his SAREPTA CONCIONE 3, write that the material of the metals should be like buttermilk before it hardens into a metallic form, and that it can be spread like butter. They call it GUR, and I have found it myself in mines where Nature has made lead. And if one is also able to make such a material here above the earth, then that should be a sure sign not only that one has the correct MATERIA, but also that one is undoubtedly on the right path. This I can make, praise be to GOD, with my own hands. When left in warmth an hour it goes into a state of Putrefaction, so that it turns black, then reddish, and finally red-brown. The philosophers call it Lac Virginis, the Milk of the Virgin. Thus, if one puts a little SALIS METALLICI in our water, it becomes like a white milk, and if one puts a lot therein, then it turns thick like butter and can be spread like fat or a similiar substance. I have thought it well to mention this, in order that you may harbor no doubts concerning the MATERIA, and this will be proven to you with the help of the only Creator. Johannes Chrysippus Fannianus, a powerful philosopher, who, I have been told, is still alive, and who is said to possess the Stone, but who has named himself such that no one will be able to recognize him, speaks as follows: "Just as the hand of the physician is required by the sick man and not by him who is healthy, so the hand of the philosopher is required by the insignificant and imperfect metals and not by the precious and perfect ones. Our MATERIA is commonly called Magnesia, which, in the Chaldean language, means 'untransformed lead', that is, such as remains just as Nature prepared it, and has not yet passed through the hands of men, much less been in fire." Hermes says: "In the metals lies the entire science, but not in the perfect ones, rather in the imperfect." RIPLEUS in AXIOMATIB. fol. 8: "Do not believe the deceivers, for our Sulphur and Mercury are only in the imperfect or incomplete metals." CLANG. BUCC. fol. 475: "If you seek a medicine which should generate all metals, you should not neglect to seek it from the Metals." ROS. fol. 379: "Saturn it is who divides my limbs. I (understand: the Sun) am the one who brings the light to the others. This occurs after the Composition, extremely slowly, which process I have obtained from my father Saturn." Here one sees clearly that gold arises in and from Saturn. Also ROS. MAG. fol. 382: "There is a plant, which is called Saturn of the Channels, or of the Reeds; from it comes our medicine." He he is speaking of the material and also of the Solution, for when Saturn is purified of all its corruption and lies in water in the final Solution, then one can place little reeds in this, into which the SAL METALLORUM shoots up like saltpeter into long reeds. One can also let it turn to salt by itself, or also turn it to a lovely shining powder. Each method is as good as the others; this will be discussed when we speak of the Solution.

MASAR SARACENUS: "Inquid; immunditia est in primo metallo", our first metal (Saturn is the first which Nature creates) contains much impurity. For this reason men despise it, and think that they cannot remove the impurity from it and that they can produce nothing with it. TURBA. fol. 154 and 155: "The Stone of the Wise is a metallic MATERIA, and an metals, both pure and impure, are within it: Sun, Noon and Mercury. Isaac Holandus in LIB MINERALI: "Ubi agit de Saturno nostro, inquit: "It is very important to observe that a metallic and Saturnian body is available, which one can easily dissolve and putrefy; and whoever knows the correct preparation may properly take great joy in it." MERCURIUS says: "You should know that our impure body is lead." TURBA, fol. 268: "I say thus of -the lead ore, that everything must come and occur from and by means of it." ROSIN. fol. 270: "I tell you that our Sulphureous living water is extracted from lead ore, and is able to produce everything. Likewise you should know that the Spirit of Silver and everything white from the lead ore must be compounded such that this LUNA becomes a white Stone (SAL METALLORUM) without any flaw or impurity." This is a fine speech, in which he says that the LUNA or LUNARIA is in the lead ore, and that the white Lunar salt should be extracted out of it. This is true, for if the whiteness which then shines is drawn out of the Saturnian body, then it is beautifully clear and one can see no LUNAN within it, but if one casts it upon a ho.t piece of sheet metal possessing the quality of Mars or Venus, then it melts like wax and the Spirit of Life flows away into the air and leaves the LUNARIAN lying beautifully behind it. This is worthy of great admiration, for it is the ALBA LUNA, of which Hermes says, that it is not to be looked upon as common LUNAM, for it is transparent. Fol. 273s "Our Water or Salt is extracted from our lead Ore." TURBA PHIL. fol. 85: "You sons of Wisdom should know that without lead no tincture can be produced, for within it is the power and virtue of the entire work." The reason for this is that the ordinary man understands everything according to the letter and thinks that this means common lead. No: for all metals are dead as soon as they enter into fire. For this reason Hermes says: "Our Stone is the sort of thing which no fire has yet touched, from which our MERCURITJS arises." We will say more of this hereafter.

In MANUALI Theophrastus calls it "Electrum minerali immaturum, electrum artificiate." It is a MASSA, compounded by means of the Art from all metals, concerning which he has written a special book. This ELECTRUM is the sort of thing in which Nature has planted the natures of all seven metals, but has left uncompleted, and thus he calls it "immature." That all metals stem from this lead has already been demonstrated sufficiently; you may read further about it in Count Bernhardus, fol. 31 and 32. I could add to these many hundreds more texts from the philosophers, but that is not necessary, for I have selected the clearest of them. He who has read these PROBATIONES and Authorities will not be able to deny that this and nothing else is the true and proper MATERIA of the Stone, since all metals arise and grow from it.

Now we wish to approach more closely and speak concerning this very first of all materials. For there have been some philosophers who, from an entire pound have obtained barely a quarter ounce which has proved suitable for the Work, as CLANGOR testifies in fol. 478. But there is a great advantage in this, for Nature has placed more Mercury or SALIS (illegible) in one material than in another, and has also cooked one material more or longer than the other. Thus, it is worth considering that more is contained in the one on which Nature has worked the longest, but that she has not yet advanced beyond her terminus, for otherwise it would be too difficult to dissolve. Flamellus speaks about this in fol. 152 and 153: "It can and must be taken first from that internal thing in which it is most concealed." This is also explained by Arnoldus de Villanova in his ROSAR. fol. 404: "There exist certain mediocre materials, among which some are more purified by Nature than others, and also more and longer cooked and digested, and these are better and more suitable for the Work." I will refer to certain of these here which I am familiar with and which I have tested, and will also make you acquainted with them, and name the places where you can obtain them. I feel that thereby I will have done enough for you, if not, indeed, too much.

I exhort you, however, my dear brother in Christ who receives this instruction from me, that you give heed to the Judgement of GOD, since you must answer to Him for any misuse of it, and that you thus keep silent about such exalted secrets. They are ARCANA. ARCA means "a chest", within which you should hold, preserve and conceal them from false men. If you do this, then dear GOD will add His blessing to you. And there can be no doubt that if the Lord supports a good undertaking, that your desire will be fulfilled, for these instructions are written harmoniously and clearly. I have no doubt concerning the need for secrecy, for otherwise GOD is not in the picture and no true faith or loyalty could be involved. You have understood enough from the cited passages of the books of the philosophers to know that they call the MATERIAM "PLUMBUM" "PLUMBAGINEM" "LYTARGYRUM PLUMBUM AETIS", and that all these are tantamount to the same thing, even though one may contain a more suitable material than the other, for the most remote or furtherest one can be employed just as well as the others, although with greater work and effort. One contains more Mercury or SAL METALLICUM than the other, but, as has been said, they all contribute to the same purpose.

The material of poorest quality is found at Sankt Joachimsthal, and is there called "glance". A centner (110 pounds) of it contains not much over three or four ounces of silver, whereas half of it is lead. In this regard you should know that the richer it is in silver, the more suitable it is for the Work, since Nature has already worked on it a great deal, and thus many tiny kernels are contained within it. It is symbolized as follows:

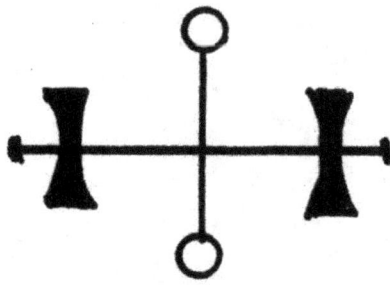

The second variety is of better quality and is found in Poland near Elkusch and also at another place thirteen miles from Krakav, where the King has his silver mine. This is better and contains more fixed silver than the first, for which reason it is more suitable for the work. It is called "lead ore" as well as "silverglance", and it is symbolized as follows,

The third variety is found near Freiberg in Meissen. When it is pure it contains many kernels of LUNAE, for the LUNA is the kernel of growth, for which reason it is called "LUNARIA". This is even better than those already mentioned if it is pure; but it is rarely found. It is symbolized as follows:

The fourth variety is even better and purer, and is secretly excavated on the Hungarian border not far from the village K'lobuck, by the villagers themselves. For the lords there strive to obtain it, as was explained to me by a villager from Klumis-Klubuck who built my houses for me. It is symbolized as follows; and there the name for it in German is "silver-lead".

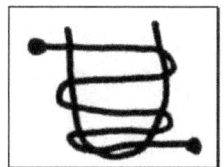

The fifth variety is fairly pure lead, but that sort which has never been in fire. It is found at Villach and is easily dissolved. For you it is clearly the best and contains much fine LUNARIAN. It is symbolized as follows; I have used it previously; it can easily be treated in the Solution.

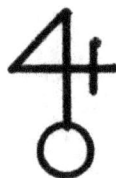

The sixth and best variety that I know of is found
in Meissen, though only rarely at this time.
However, if one orders it, one can still obtain it,
and it is called "GLASSERTZ". It can be cut and
stamped like lead, and a centner of it contains as
much as 12 to 13 ounces of silver. I have obtained
11 ounces of Mercury out of one pound of it, and
although CLANGOR writes that he only obtained a
quarter ounce from one pound, he must have had the
poorest quality of lead ore. There is also GLASSERTZ
to be found in St. Arinaberg, which is also rich in
silver, but it cannot be cut and stamped. (However,
there one can also find some that can be cut and
stamped and is not distinguishable from other lead.)
This I symbolize as follows; It can also be
dissolved, which is the case as well with that
variety which is indistinguishable from common lead.

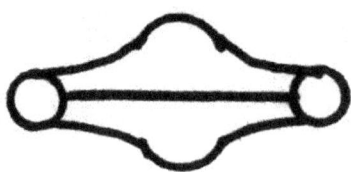

So, my dear friend, here you have an instruction not only in the general agreement of the philosophers, but also in the materials. May it please the Eternal and Almighty GOD that you use it to His honor and for the benefit of your neighbor.

How one should proceed further will be discussed in the second part. This treasure is not to be purchased with any gold, as Soloman in his wisdom says in chapter 7; and silver is as worthless sand when calculated against it.

There may well be other things concerning the GUR and the PRIMA MATERIA which are alleged in other sources, but there should be absolutely no doubt about this; for I have surely seen and experienced so much that I could demonstrate them myself (if they could be the case), but they cannot be the case. Thus you, my good friend, possess in full the entire tractatus of the whole material on which all philosophers agree.

It could have been easily less long, but because I did not know whether the HAERES knew Latin or not, I have written many things twice, although with greater trouble. My son, merely refer to the books of the philosophers; you will not find it different then I have said; for I can now demonstrate the entire operation in practice, praise be to GOD. In this regard the best proof is that the philosophers are in concord and agreement with one another, for the truth is contained only in the metallic root, as Count Trevisanus testifies in fol. 14 and 16. ROSAR. fol. 36, FLAM. fol. 147, REUCHL. D. VERB. MINF. fol. 100 say that concord or agreement is an obvious proof of the truth.

But on the other hand, the foundation of lies is discord or equivocation, and here I will let the matter rest, for he who is not content with this proof will never be able to be helped.

As soon as the other points are discussed, then everything will be illuminated, and the play of miracle may then be seen. But I ask and protest in the name of GOD, that no one brings this tractatus before the eyes of another. Whatever has not been understood I will attempt hereafter to explain as completely as possible, but do not take your questions to another man, but rather pray diligently. May honor, praise and glory be with Him, who lives from Eternity to Eternity, AMEN. Hold GOD before your eyes, pray and read and work: GOD will help you as He has helped me.

SOLI DEO GLORIA

(The End of the Greater Edifyer)

SECOND PART

THE LESSER EDIFYER
LILIUM INTER SPINAS

Rex Salomon Sapientiarum capite leptimo Optanti mihi et oranti prodentia data est, et sapientiae spiritus obvenit quem ego skeptris et foliis antepono et dMtitas, si cum ea conferaritur, nihili facio, neque pretiosas gemmas cum ea comparandas duco.

Item omne (Sol) cum ea collatum ferme est arena, et prae ea (Luna) proluto ducendum.

Hanc ipsarn supra valetudinen et forma.in amo, earn habere malo, quam lucem quippe cuius splender sit inextinctus.

Cum ea autern mihi cuncta simul bona et eius opera immensa dMrae evenerunt. Nam in exhaustus hominibus thesaurus est, quo qui utuntur, ii cum Deo ineunt admicitiam ob doctrinae dotes commendati. Est enim vapor quidam dMnae potentiae, et sincera comniptentis effluentia,..

Poet Nubila Phoebus

There was once an old adage or proverb that went, "after great suffering there usually comes great joy." Sic et contra. Such was also the case for me, unfortunately, a few years ago. There are no doubt many others as well, who, when they begin at first without a true basis or foundation, end up approximately in the condition that I will describe at length. For, because I thought that I held the whole world in my hands, I succeeded in less than nothing. Because the glass in which I had set my well-being broke with a great crash, the material besplattered my MUTOS PHILOSOPHOS and books now and again to my great damage and detriment, concerning which I will forebear to communicate further, except to say that I was very be-numbed, stupified and frightened by this unanticipated setback, even to the extent that in my sorrow and affliction I did not know where I was, what I had intended, nor what I should do next, for all my joy, delight and bliss had been transformed and transmutated into vain poison and bitter gall, instead of into gold and silver (as I had hoped). After I had come a little to myself, I first considered my great damage and detriment, and began, with hot tears, bended knees and heartfelt sighs, to lament and complain about this to Him who lives from Eternity to Eternity (for GOD gives and takes as it well pleases Him, and to whomsoever He will), earnestly praying that He might yet now befriend and have mercy on me, that it might be possible that He should lead, guide and show me the correct path to the mirror of His Majesty through the Spirit of Truth and Wisdom. I consoled myself in this through the thought, that Dionysus and Zacharias had said that most of the philosophers had also erred greatly in the beginning, and nevertheless had finally come despite this to a happy and fruitful end.

After I had troubled myself at length with gloomy thoughts, I was struck with a question of doubt that went beyond them: Whether Almighty GOD would even allow us poor sinners, who now live in these last and evil times, to have, share in and know such an exalted Secret. After much thought and emotion, I came at last to the conclusion that even those who had possessed this Mystery before us had themselves been sinners, and that they attained and received this, not from merit, but from grace. Therefore even now one who behaved piously and held GOD before his eyes could, through the grace of the loving GOD, attain it and grow powerful as well as they. Thus, I once again seized courage and valiantly took up my philosophers, magi and MUTOS REGES, especially Count von Tervis. For although I had read him before many times thoroughly, nevertheless I could not find a sure foundation in him. But now, since the hour of revelation was at hand, I came in my reading to the passage in which he describes the MATERIA, and suddenly it seemed as if a fiery spark of CALIBI lead me from that place to the passages in which the power of the Work lay. I was frightened at first, but then, as I looked further, the eyes of my understanding were opened, and I could see and understand that to which I had previously been blind and for which I had yearned so long. Then I felt joy in my heart, and thanked GOD, requesting additionally that He might instruct me further in the means whereby I might attain to the completion of this exalted Work. Thus, I soon undertook to journey after the materials (although they could be found everywhere). For I wished to obtain the PROXIMIOREM or PROPINQUIOREM and not the REMOTIOREM, since the one is richer than the other, whereby it better achieves the goal, as is explained by GEORG. RIPLEUS in his AXIOMATIB. , 12 PORTATUM, and also by FLAMELLUS in fol. 126, 150 in fine, in these words: "Hoc vero imprimis occult issimum est, ex qua re minerali neri debeat propinquis."

As I was now on the way and my heart was full of thoughts, it happened that I met a venerable old peasant between two mountains. He was clothed in a long gray cloak or smock, on his hat he had a black

ribbon, around his neck a white banner, a yellow
belt about his body, and also red boots on his feet.

After I had greeted him and approached him more
closely, I became aware that in his hand he held two
star-like flowers, each with seven rays. The one
white and the other red, which he was contemplating.
They were very beautiful and radiant in colour,
lovely in scent and sweet to the taste. Also the one
was feminine and the other masculine, and yet both
grew out of one root and under the influence of all
the planets. I asked the peasant what he meant to do
with these flowers, for, although I knew them both
well, I did not know that they had an OPININEM
DISTINCTAM as man and woman, that is, that they were
of two distinct natures. He looked at me earnestly,
asking who had led me to this unusual place, for it
was sought by the most exalted men of the world, but
it was closed and barred to them. However, after I
had told him of the wonderful course of my life, of
which I have already mentioned something, he
laughed and, turning to me in a friendly manner, he
said: "You should know that no one may attain to the
knowledge of these flowers unless he is predestined
for it, or unless he achieves it by means of
diligent prayer and strong, firm faith, and even
then it is not given to him without great effort,
trouble and tribulation, as you yourself must

89

confess. And this is so in order that those who possess it remember this, and learn to esteem this Mystery more highly and to keep it secret. But since you have now come so far, I will show you, with the permission and licence of the DIVINA NUMINIS, that from these two flowers comes the PRIMA MATERIA of all metals, but only after their Conjunction, and not before it. Concerning this read Count Bernhardus in fol. 45. Almost at the end of the second part of the same book he calls these two flowers a red man and a white woman. However, because of the dangers involved, the philosophers have always written of the PRIMA MATERIA, in order to conceal its root from the unwise, and have remained completely silent about the SECUNDAE MATERIAE. For you must first obtain the SECUNDAM NiATERIAM, which is crude and itself the SUBJECTUM LAPIDIS, and must extract from it the man and the woman, which only after their conjunction turns into the PRIMA MATERIA, which I thus reveal truthfully to you."

I was astonished at this speech and was accordingly very glad that in many points it was in agreement with my own opinion. When he had finished, I said to him: "My friend, I truthfully would not have sought such exalted wisdom from you, for you appear so plain and simple." He smiled and said, "For precisely this reason the entire would errs and lacks my wisdom, for my externally insignificant figure usually fools them. But if they would only ask in friendliness to be allowed to take off my old gray smock, then underneath it they would find, as you now know well, a shining diamond suit of armor with a ruby lining. However, the Most High has sealed up all such things from the many, in order that they will not be able to contemplate that from which all the metals have their origin."

I replied to this, "My dear and gentle peasant, these flowers have a glorious radiance; are they not therefore also medicinal?" He said, "They are indeed medicinal; but their great power properly lies hidden in them. For when they are still contained in the root they are very poisonous, for which reason the root must first be sublimated very gradually and gently. (You already know the Sublimation of the philosophers, for otherwise I would have told you about it). This must be done without an extraneous or acidic agent, which could corrupt their growing power and nature, for otherwise they are of no use. These two glorious flowers grow variously, without the addition of other things, out of this poisonous mountain, and had I myself not known (he said) under which planets these Edifyers were constellated, then I would not have arrived at this miraculous place, and thus would not have come to know such a secret. In this you can believe me truly."

I spoke further. "My dear friend, you have told me many strange things, but please tell me also whether these two flowers grow simultaneously together, or how it is that they have but one form, for I deem that the greatest Art lies in this one point, (although I also consider the Resolution very important), because the philosophers do not say very much about it." As I said this, the peasant shook his head, and then remained silent for a while, but finally began to speak. "It seems to me that you are very curious about these matters (although I cannot blame you for this), but allow yourself to be satisfied for now with my manifold account, for this is indeed the cornerstone, upon which most men stumble, for there are many of them who know the true MATERIA, but to whom this technique is concealed. However, come again tomorrow at this time; I will again be here, and reveal, teach and make known to you as much as has been given me."

I thanked him for his true report exceedingly, left him joyfully and awaited the following day with great longing. I did not tarry, and arrived punctually at the place. The peasant was already there, holding the two flowers in his hand. I wished

him a good day and reminded him of his promise of
the day before, but with veneration, saying that I
wished to know how to be of service to him and to
please him in everything. He replied that he
remembered his promise well, and desired to fulfill
the same, but that I could not be of service to him
in any respect, but rather I should know that if I
were on good terms with GOD then he would be my
friend; if not, then he would be my enemy. For he
also must follow and obey the commandments of the
Highest, and I should be content with this. He spoke
further: "Listen, you child of man, for now I will
repeat my speech in short from the beginning, and
fully report to you and teach you the entire process
with all its requisites and details. However, you
must pay strict attention to it, and consider my
words again at home often after having prayed, for
otherwise my meaning will be concealed unknown in
your senses, and lead you upon void and erring ways.
For it is a gift from Almighty GOD; believe this
with certainty, and now listen. Let us set ourselves
upon this green, for I am old and cold by nature,
and also have an infirmity in my legs, so that I
cannot stand for long. I would like, therefore, to
rest on the green.

"You have doubtlessly read that our Magi,
philosophers and kings write and speak in accord

with Nature; know from this that he who will achieve something in this Art must first know and understand the origin, birth, distinction, friendship and enmity of all metals. And perceive further that all metals grow from one root, and that their original material is of one kind, but that they are distinguished only in that one of them is purer than the other, and also more cooked and digested. All of this is written and demonstrated in the books of all philosophers, in which alone the truth is found, and not in the falsely written recipes and processes of the traveling scholars. This you may believe for certain. He who would know the ground and fundament of these things should not allow himself to be deterred from the reading of these books. Nam qui vult sentire commodum, oportet ut etiam sentiat onus" - it would take me far too long to tell you about all of them, and when I have finished fully informing you, you will not see me again or speak to me until you have brought the Work to its conclusion. At that time I will talk with you in a much more friendly mariner, and you will then also know me better, love me more strongly and hold me in higher esteem as now, and we shall thereafter not soon part-but enough of that for now.

"You should also be informed further, that he who understands the origin of the metals will also know

well that the material of our Stone must also be
metallic, but that it is not a metal, nor a mineral
either, but rather metals and mineral~, minerals and
metals. For the nature and character of all of this
is in one thing, which is called ELECTRUM MINERALE
IMMATURUM MAGNESIA or LUNARIA, for which reason the
philosophers always use the plural form, saying
"METALLA, METALLORUM, METALLIS," etc. I must not
speak about this more clearly, neither is it
necessary, because this material is already known to
you, and otherwise others may overhear it. For in
these mountains there lurk very many temptors, some
of whom come very close to us, and a few even all
the way. Thus I do not allow myself easily to be
seen with these flowers, for so I am commanded. As I
have said, this mineral root must gradually be
separated from the corruption that it has received
from the poisonous vapors, and then the white
Mercuris lily-sap must be pressed out of it. This is
very delicate and volatile, and is thus to be sought
in the upper parts of the flower, and its name is
AZOTH or GLUTEN AQUILAE. You should also not neglect
to seek the Sulphuric, incombustible, fixed, red
lily-sap in the lower parts. This is called LATON or
LEO RUBEUS. There have my explanation, according
to your desire. Do not ask further, for I am
restricted, and forbidden to explain more; Pray
diligently about it and it will be given to you. It

95

is especially astonishing that these flowers never wither or wilt, and that the one can be transformed into all sorts of forms and natures, and also loves all the planets to which it may be united, such that it may never again be separated from the house of the planet to which it is united for all eternity. To describe the virtue, nature and quality of this flower is beyond the heart, mind and soul of any man, as all wise men must confess. Now, as you can see, these two flowers rest upon a seven-fold stalk of many colours, but they have spread quite far apart from one another, because of their differing natures. For this reason, one must find a means whereby they grow together, so that, from both of them a glorious, incorruptible and eternally enduring fruit may break forth, sprout and grow, though this cannot occur without the will of GOD. Furthermore, you should know that the number of the white lily seed is very different from that of the red, which the wise have concealed very thoroughly, calling it their PONDUS or weight. Without this knowledge, the two lilies cannot be united, nor be mixed PER MINIMA together. The ancient Arabians also speak of this: "Pondus masculi singulare et forminae plurale semper esto." This is explained by the Count, in that he says: "Terrena potentia super sibi resistens et pro resistentia dilata est actioagentis in altera materia." Do you understand this?

I answered that it was somewhat obscure. He said "Do not trouble yourself too greatly about it, for when you attain to the growth of these two lilies, then you will see yourself, from the quality and nature of each of them, what you should do. But use a gradual, gentle warmth, for otherwise the seed of the white lily will smoke forth out of it in the form of a vapor, and all your effort and work will be in vain."

I spoke further: "You have always mentioned only these two lilies; however, the philosophers also occasionally speak of one thing alone, such as that everything sought by the wise is contained in Mercury or in AZOTH, and they also talk about three things, namely Salt, Sulphur and Mercury. But mostly they refer to body, soul and spirit, and you have not mentioned these at all."

"It amuses me," he said, "that you do not yet understand the TERMINOS PHILOSOPHICOS, or perhaps you wish to tempt me; but nevertheless, I will enlighten you concerning this. When they speak of one thing, then it is the SAL METALLORUM, the LAPIS PHILOSOPHORUM.

I have spoken here of two things, that is, CORPUS
and ANIMA. The third is the COPULA AMBORUM, that is
the SPIRITUS, which you cannot see, but which is
nonetheless concealed in both of them, and therefore
also hovers above the waters, as you may read in
Genesis 1. Content yourself with this; however, I
will continue to speak of the two. Now, when these
two lilies are polished very purely, enclose them in
a crystal very securely, without fire, set them in a
gentle sweatbath, and soon the white lily will open
wide and the red will contract and close. But
because the red is of a fiery nature, and finds help
from the external warmth, this warmth releases a hot
balsam scent from it into the coldness of the white,
so that they become disunited. For neither will
accede or give way to the other because of their
contrary qualities. As you know well, both grow as
high as the heavens, but are again driven downward
by the winds, and this repeats itself several times,
until they must rest on the earth, faint, tired and
sluggish from the labor of their rising and falling.
This know that if the bath is not regulated
sufficiently, so that their two natures do not rise
simultaneously, but rather only one at a time, then
you will never enjoy or share in their scent. For
this reason, pay full attention to this first
operation. For when these two enemies sense and
perceive that neither can have an advantage over the

other, then they become united, and with such love and friendship that from then on they desire to remain with each other eternally. In this union the entire firmament comes into motion, as well as the Sun and the Moon, to the extent that they both become darkened, as long as this pleases the Most High.

"In accord with this GOD in His love created his bow of many colors in the air, as a sign that you should not doubt; for GOD will be merciful and will not permit the drowning or flooding of these two. You may properly take joy in this, for in a short time you will see that the Moon gradually breaks forth again, and is not as dark as before. Finally it will shine again, with a pure white radiance, and will be beautifully clear. However, the Sun will still be hidden behind the Moon, and you will not be able to see it because of the earth. But if you have clear eyes of understanding, then you will perceive four planets in the Moon, which you can transform and transmute into their enduring nature through the Moon's radiance. But when Sirius, the Dog Star, approaches near the Sun, and the heat grows ever greater and more intense, then the Moon will be darkened by the radiance of the Sun, until it is finally concealed behind that radiance. The Sun will grow intensely angry, due to the impurity of the other planets, and through its wrath will turn first yellowish and finally blood-red. But because they humble themselves before him as their lord (since GOD has so ordained it) he will again favor them with his pardon, and make them all similar to him, so that they may always remain with him in his kingdom and reign forever. For this they may properly have great joy, thanking and praising him, and also the Most High, through whose permission they have been so gloriously adorned, and hoping daily to use such adornment in the praise of His great name. See, now I have freed you from your doubt, and hope that you will now understand this matter perfectly, and thank GOD, the Creator of you and me, for it, and well know how to set forth and advance this exalted Work at home. But pray diligently, and use it correctly, or you shall never see me or find me again."

In my joy I truly didn't know what I should reply to the peasant, but was nonetheless sincerely grateful, and asked him further in a friendly manner whether there was anything more than this to be done, and whether the Art could thus be completed. He answered me very gently: "You should know that the virtue of these two lilies can be renewed and propagated every three days, that they may increase in their power and seed themselves one thousand times, and that this occurs when the seed is planted and sown in the prepared earth. Thus, on the first day the darkness shall occur, on the second day the bright moonlight shall come, and on the third day the red setting sun shall break forth again. This Work may proceed as long as it pleases the Most High. The jewels and pearls of these flowers also grow forth out of Nature, but the highest thing is that which pertains to you men in regard to a further knowledge of GOD and a long life. For if someone should partake of only a petal of this, he shall soon recover from all sickness and disease. And although I cannot tell you now of its magic power and other exalted secrets, when I come to you again after the completion of your work, then I shall relate to you and give you to receive many of its higher virtues, qualities and circumstances. But for now you should content yourself with this and contemplate my words well. With GOD'S permission all your desires shall be granted you, and since I have now beneficially fulfilled my duty, I must now again depart from you. Yet you shall remember me and await my arrival after the passing of several moons."

"Oh my dear brother and best friend," I said, "you speak so excellently and sensibly about these exalted matters that it truthfully seems to me that you cannot really be a simple peasant, although you appear to be this externally. Furthermore, you also speak Latin wonderfully, and this I am not accustomed to find among peasants. So tell me, please, in which university you have learned all this; for you are clearly of a very high rank, such that I have never encountered your equal." At this the peasant began to chuckle, and asked me what moved me to such a question. I replied, "I wish very greatly to know whether you have learned this in the universities, for there men think that they possess the true philosophy."

"I am amazed," he said, "at what you allow yourself to dream. If you intend to seek or find philosophy or truth among those who themselves despise it, verily you cannot. But in regard to my wisdom, this I have received solely from Him through Whose word and command the heaven, firmament and earth must tremble and shake. Accordingly, I have already told you that inwardly I am adorned with gold, diamonds, emeralds and rubies, and have only wrapped this gray smock about me in order to hide and conceal myself from the powerful, for they hope to catch me and gain power over me."

I asked further, "Why do the great lords and potentates not seek this philosophy from their philosophers, whom they must retain in universities each year with ever-increasing wages?" He answered, "They are not worth wasting many words on, because they do not allow themselves to be instructed and taught, and for this reason, will not be able to endure for very long with their useless and futile arguments and invalid reasons, GOD is very angry with them, because they mislead the youths with their sophistical matters. They teach and dispute always DE LANA CAPRINA and when the disputation comes to an end, one is always as clever as the next, and all have attained the same thing, namely a great fog. Therefore guard yourself against their vain poison-loving phantasies and allegations. They chatter a great deal concerning the nut and do not realize that the kernel within it is what is best. They teach the youths the ARTES DICENDI and drill them also in the PITIUS GRAMMATICUM, but all the rest is only bellow-mongering. If they were not doctors or masters they would probably still practice the true philosophy. But now they are ashamed that as graduated persons they should still have to learn.

Therefore the true philosophy must be innocently cursed with the appearance of falsity, and be persecuted and slandered to the utmost, but this must occur so that the wisdom of GOD is called fool-ishness by the world, and vice versa. "Narn devs

nori sine gravi iudicio sapientum sub nomine
stultitiae voluit esse revelatan ut nimirum
mysterium virtutis suae effet arcanum: sed tandem
bona causa nostra triumphabit.' And we esteem them
far less than they think, and can do without them or
advise them far more easily than they can do without
or counsel us."

I said "My dear peasant, I must also confess to you
that I am of the same opinion, for, in regard to the
precious truth, I must say that I have learned much
more outside the university than through the pseudo-
philosophers, who have also been repugnant to me,
because their doctrines are too trivial." For me,
"said the peasant, "that is enough of that, and
besides, it is soon time for me to go to my previous
place." "Please do not leave me now my dear
brother," I said, "but rather grant my friendly
request for one more question, so that I may master
this doctrine. Then I will gladly be content,
although I am already so sincerely thankful for
everything, for then it can be mine.

"I don't know," he said, "it may be the sort of
question that I can allow you, but it might also be
something that I am forbidden to reveal to you now;
but speak on." "The wise philosophers," I said, "all
write that the greatest Art resides in the mastery

of the Fire, especially since this should be kept uneven. And then I would also like to know, what should be the PROPINQUIOR or first material of the Stone, from which I should or could extract the FORMAM SPECIFICAM or the two flowers; for, although I know a MATERIAM GENERALEM I am nonetheless uncertain about the former. For CLANGOR writes that barely a quarter ounce of what is suitable for the Work can be extracted from an entire pound, which would be very little; however, I consider it possible that several ounces should be preparable from one pound: both of the white and the red."

"I see well how the matter lies;" he replied, "you want to know too much, and not to seek and work yourself. No, my brother, nothing will come of that. It is called the LILIUM INTER SPINAS (Lily among Thorns), and if someone tries to pluck it so easily, he will only cut his hands. For this reason one must first gradually remove the thorns with industry and labour, and then proceed delicately and cleanly to the glorious lilies, in order that one may finally enjoy them. Besides, you have mixed two questions together, but you have only asked for one answer. Nevertheless, I may indeed tell you that you should consider well the four parts of the year and should distribute the parts of the Work accordingly. The books of the wise discuss this sufficiently. Thus,

you perceive well that it is hotter in the dog days
than in the Spring, and colder in Winter than in
Summer. At this point many might be clever and say
that even children know this. But, my companion, you
do not yet know what Winter and Summer are (for the
Philosophers). So don't proceed too quickly toward
it; you have sufficient time to attain it. I can
tell you no more about it, but since you must fail
in it at first (which will nevertheless be a great
blow for you), I will now teach you a method by
which you can abundantly restore your losses, and be
able in a little while to have and achieve your
nourishment. See, under my gray cloak I have a green
lining. If you polish it with flint and iron rust
and the fixed Red Eagle, then my green lining will
become much more glorious. This you should immerse
in the pure Moonlight, and the Moon will borrow and
receive three ounces from the Sun, and give and
impart these to you for your nourishment. This you
may enjoy almost every eighth day, which you should
calculate. Even a large man, who needs a lot of
nourishment, can nurture and preserve himself
abundantly with it, and the cost is not great. This
you should keep secret and thank GOD for it. And
now, farewell."

And when he had said this to me, the peasant leapt
hurriedly and quickly disappeared into the mountain,

but the two flowers remained, and stood on the place where the peasant had vanished, I hastened toward them in order to pluck them, but they swayed to and fro, eluding my hand. And when I snatched at them quickly to catch them, behold, there lay a piece of the proper, crude and true MATERIA LAPIDIS, weighing several pounds, before me in their place. Then a voice came forth from the mountain, saying: "Deus sua bona vendit laboribus." After this I heard and saw nothing more. I then fell down upon my knees, thanking and praising Him Who lives from Eternity to Eternity, Who is Himself Wisdom, praying that He might fill and enlighten my heart, mind and soul with the Spirit of Wisdom, that I might gain and receive such a precious, exalted and worthy treasure, promising and swearing further that I would use and employ this in honor of His holy name, for the benefit of the Christian Church and especially for the highest good of my neighbor and of the blessed poor. And thus now you as well, my dear ones, have received in abundance the true foundation of the highest, most precious and worthy treasure, with all of its details. Do with it as I have, be of good enterprise; avoid the sophists and hold GOD before your eyes, and you shall not labor so often in vain, but rather see and sense perceptibly the moracle of GOD, whose name be praised and blessed from Eternity to Eternity, Amen.

Petite et dabitur vobis

Feliciter absolutum in arce - - (Sol) axhaffz Anno
aere salutis 1598. Julii 9. stylo veteri.

SECOND SECTION

(of the Lesser Edifyer)

After I had thanked the eternal Almighty GOD and
Creator of all things for his gracious revelation,
and had heartily praised and honored Him, I took up
my MATERIAM SECUNDAM (PRIMA MATERIA will be
discussed hereafter) and kissed it joyfully, as that
for which I had had such a heartfelt longing and
desire, and for whose sake I had been for so many
years in doubt, affliction, sorrow and care. I
looked at it with great astonishment, especially
because it had no particular external appearance,
and yet should be able to bring about and complete
such an exalted, important supernatural work. Then I
remembered that the peasant had said that GOD had
provided it with such an appearance for highly
important reasons, and, in particular, so that the
poor might possess it as well as the rich, that they
could not complain or lament to GOD that He had
given the rich an advantage in this respect. For
truly the rich pay no heed to it and believe even
less that such a power is contained in it, as can be
read in ROSARIO MAGNO, fol. 248: "Si materiam
nostrani homine nuncuparemus, insipientes et divites

eam esse non crederent." Thus the poor obtain it sooner than the rich. After I had wrapped my material up and preserved it well, I proceeded homewards with joy and sang on the way the following song from chapter 7 of SAB. SALON: "Where is there a GOD like our GOD, who can give help in every need; I was in sorrow, now I am full of joy." When I had arrived home, I then set forth to prepare and make a goodly supply of SUMPTUM from the PARTICULAR, as the kind peasant had taught me, so that I could await the UNIVERSAL with all the more peace and steadfastness, beginning thus in the name of GOD. I bought a good supply of coal (for the process consumed much coal), and built a serviceable furnace and oven for this, and thus in a short time (a few weeks) had prepared my good supply.

This the Enemy of the Christians could not abide, and incited alarm among the people in all the streets. My neighbors complained to me that I would burn their houses down; my friends and other acquaintances reproached me by saying that there were many counterfeit coins now current, and available, and that I should refrain from such futile things and practices, in order that I should not come under suspicion, but rather pursue and safeguard in earnest my FACULTATEM JURIDICAM (since I was a doctor of law), for thus I would be able to preserve my abundant well-being. But that I could earn my bread in this way with a good conscience I doubted, unfortunately, very much. What next occurred was that all the smiths and goldworkers complained before the Council that I had raised the price of coal, and thus that they could not have their handwork and rightful nourishment, and PER CONSEQUENTIAM were not able to give the Council and the city any taxes or tribute. For I had paid more for the coal so that I could get it before them. They pursued this to the degree that, the Council forbade me to continue and ordered me to stop purchasing coal and to follow my occupation. In all, the consequences were such that I had to tear down my oven and then depart, in order to seek a good friend who could provide me with enough money that I could more peacefully await the UNIVERSAL. (Nevertheless, I told no one what I had in mind.) The tribulation connected with travels and other inconveniences lasted almost into the third year. GOD knows how I took it to heart that I knew it all and yet could not take up my work; I thought again and again that GOD perhaps did not will it and would not permit it: "nam quo nos fata trahunt retrahunt que, eo nos sequi oportet." As the Count von Tarvis also relates, he knew the entire science of the UNIVERSAL perfectly for two years, and yet, because of many impediments and hindrances, could not set out or proceed in his work, all the while entrusting everything to the loving GOD. During my travels I conferred with learned men and grew ever more understanding; I persued MUTUAS OPERAS in the Arts and Sciences, as was then the practice. I also collected lovely materials of all sorts of ores and stones. But I found very few, indeed, not more than three, who were travelling on the right philosophical path. For they all wished to deal with common mercury, gold, antimony glance, cinnabar ore, and also many more insignificant futile things, in which they all erred greatly, because they did not follow and work in accord with Nature. If they had acted in concert with Nature and had brought the correct things together, then they would not have erred so lamentably. However, such an exalted gift is not given to everyone, and accordingly each man must make his own reckoning and test himself well, before injury overtakes him and harms him: let him heed who can.

After I had thus completed my CURSUM ITINETUM SECUNDUM FATALEM CONSTITUTIONEM, I came cheerfully back home, where my supposed friends again appeared, and wanted to know where I had been so long, what I had accomplished and what I wished to do with it next. I answered them shortly, saying: "Is the world not large enough? You think, perhaps, that your city is the entire world and that a man cannot provide for himself without it. Oh no, but if you had also sought after something, then you would judge of this far differently. But there are, praise be to GOD, enough other people who recognize and accept with great thankfulness that which you have despised and ridiculed. You should know, by the way, that I will not be troubling you much with purchases of coal, for I no longer have any need for it." They were greatly astonished at my speech and put their heads together, but they did not know where the rabbit lay buried. I, however, completely ignored them, rented a house, hired a helper, and took up the Resolution with great desire, after praying and gMng thanks to the Eternal GOD. I did not stop until I had completed and perfected this process, which is perhaps the most excellent and difficult part of the entire Work. "Nam hic iacet multa tarditas," as the philosophers write, "et est davis artis." This can easily be harmed, destroyed and burnt by the fire, so that the flowers or the power of growing, dries out or burns away, for which reason I had to use great caution, and take good care that I received no injury from its corruption, as Theophrastus relates in his MANUALI. However, it all turned out well for me, through GOD'S will. When the poisonous vapors had departed from the Stone, then our two flowers, of which the peasant had spoken, very gradually came into view. However, I perceived the white one earlier, for the red one was not so highly developed and grown. I took a petal of the white one and tasted it. There I truly experienced a very sweet, glorious and lovely flavor, to which I had never tasted anything similar, and for which I felt sincere joy. The remaining petals I laid upon a hot piece of sheet metal, where they swiftly flowed away and smoked IN CONTINENTI. Through this I recognized that this was the female, because it was so fleeting and volatile. Thereafter I needed great care to master the red lily, which pays no heed at all to any fire, but rather reigns constantly over it.

Nevertheless, before I obtained these two lilies, I had experienced a great deal of unpleasantness, which I will not relate here. This was soon forgotten as soon as I had obtained the two lilies. I thought about the peasant, and was astonished at his high and genial understanding. I followed his instructions further, and joined the two lilies gently and delicately together. Here I perceive a remarkable wonder. Next I enclosed them both in a crystal vessel, without great warmth, gently and gradually. When the sun began to shine, the white lily opened out completely, as though it were merely pure, clear water, like the morning dew that lies on the grass, or like a delicate, very bright teardrop, like clear moonlight, yet with a faint bluish reflection. As I watched closely, behold, it took the red flower into itself and swallowed it, so that I could no longer see even one of its petals. However, it could not conceal it very long, for the red lily is of a hot and dry complexion, whereas the white one is of a cold and moist nature; and since the external sunshine came to the aid of the red lily, it endeavoured again to come forth, although it was still unable to do so because of the strength of the white one, whose nature had predominated up to then. Nevertheless, they struggled mightily; both grew simultaneously up to the heavens, but were again driven downward by the whirl-wind. This continued until both of them were forced to remain joined together below (since they had not taken heed of the growing root). With this they became the PRIMA MATERIA LAPIDIS ET METALLORUM. After this darkness gradually began to be manifest, and the Sun and the Moon became ever more blackly dimmed. This lasted a considerable time, as can be read in the Count. But after a good while the sign of grace appeared to me; the rainbow of many colors, of which the peasant had said that it would be a sign of joy and a promise of a good end. Then, when a little of the moonlight could be seen, although it was not yet very bright, the Sun began to shine more hotly, until the Moon grew full and shined forth clear, and transparent, as though it were a transparent pearl or a crudely cut diamond. The four planets took great joy in this, for by this means they could be transformed from all their corruption in the radiance and nature of the Moon. In his parable the Count calls this the Kings Shirt. I now applied the third grade of fire, and soon there grew forth all sorts of glorious fruits; quinces, lemons, sour-oranges, and apples, all lovely to the sight, out of a hyacinthine earth. Within a short time these had transformed themselves into lovely red paradise-apples, growing out of a ruby earth. Finally they changed and coagulated into a glorious, bright and evershining body, which made all the darkened and dimly colored planets brilliant and shining with its own radiance. All this took a very short time.

After I had made several PROJECTIONES with many pounds of purged metal and had delighted in the amount of it, astonished that such an insignificant, indeed paltry, quantity of our Stone, should have such great power to penetrate and to transform all sorts of metals (namely in a ratio of one part to one thousand parts), I sat down again to work, after having prayed and offered thanks, and was ready to make one more PROJECTION to see if I could come closer to the foundation of the PROJECTION; but as I was about to begin, suddenly my dear old peasant came again to me through the locked door, greeting me cheerfully. At first I was shocked, for I didn't recognize him at once, and besides, he appeared very suddenly before me and was clothed this time in a plad coat of many colors. I slumped down on the bench, because my legs were trembling. He spoke with a laughing mouth and friendly gesture: "Don't be afraid, my dearest brother, for you have a merciful GOD, and, in addition, that which your heart has desired on this earth I have come again 'to you, as I had sworn and promised, in order further to reveal and to instruct you fundamentally in these, and in other even higher and more wonderful things and secrets (for this is only the beginning). Therefore, learn now that it is an insignificant, simple and easy thing, as you must now yourself confess, to make the Stone, as GOD has ordained for highly important reasons; but to understand the same properly and perfectly is such a task that all philosophers, even Adam, Soloman and Hermes, as well as Theophrastus, even if they be the wisest of all, must bow down, lower their heads and openly confess their inability in this regard. Zacharias, who often made the Stone, also attests publically to this in FOL. SEPTUAGESIMO PRIMO, saying: "Nostrae medicinae scientiatam diuma tamque, supernaturalis in secunda quidem operatione, ut semper fuerit ac sit adhuc impossibile hanc innotescere hominibus quocunque studio vel industria quamvis omnium etiam sapientissimi sint atque doctissimi Philosophi, nisi primum a Deo sint inspirati, deficiunt enimhac in parte omnis naturalis ratio et experientia." That is, our Art and Science is so divine and supernatural (understand; after the Composition) that it has never been possible to understand through which means it could or might be able to exist, even by those who have been or still are the wisest of the wise, unless they have been previously enlightened by GOD. For in this point all of our sense and natural reason shatters, However, in order that you may be further introduced to and instructed in this, as I have promised, I will teach you thoroughly and inform you of as much as is granted and permitted me now to disclose and reveal. You may then appeal further in accord with my guidance, most diligently to the Almighty and Most High with fervent prayer, for from Him come all treasures of wisdom, At that time, without doubt, you will be enlightened and gifted with high understanding and extreme wisdom, as the highly wise King Soloman testifies in chapters 7 and 8 of his Book of Wisdom. For the Eternal GOD wants properly to be asked for this, and then gives it gladly (as He has given it to others previously) to those who long for it in their hearts and who intend to use it in honor of his blessedness to help the pious, their neighbor and the blessed poor.

However, because I have perceived that you have carried out the projection or casting of the tincture unwisely, you should be instructed that you must thoroughly purge and purify the metals of their ADJUNCTIS and ADHAERENTIBUS ACCIDENTIS, or Sulphuric impurities, before you make the projection, or you will suffer damages. But as to how such purification should proceed, this you have read in the books of the philosophers, and it occurs thus: .... And as he said this, he picked up a piece of copper, placed it in a crucible, added some purging powder to it, and, with a bent iron wire, withdrew several times its stinking, red, combustible, corrupt Sulphur like dross from it, for it had kept the tincture from penetrating, until the VENUS grew pure and the dross turned whitish. And when I then projected my tincture upon it, this penetrated and entred it in an instant, so that the entire CORPUS VENERIS then turned to glorious, better than natural Hungarian gold. I was delighted in this and thanked him for his true instruction. Thereafter, he told me about the purgation and purification of other metals, which were a delight and refreshment to investigate.

He also instructed me further, saying: "Furthermore, you should know and understand that with the white fixed Stone you can make all sorts of precious jewels that have a white lustre, such as diamonds, white sapphires, emeralds, pearls and the like. With the yellow Stone, before it turns deep red, you can prepare all sorts of yellowish jewels, such as hyacinths, yellow diamonds and topaz. With the red one, you can prepare carbuncle stones, rubies, garnets and SPINETEN. All of these will far exceed oriental gems in preciousness, virtue and gloriousness. I will show all of this to you later and demonstrate it with my own hand, for it can easily be done. But now I will first let you see a wonderful mystery, although you must first shut all the windows, and not be frightened by it, but rather delight in its exalted nature and power, which are implanted in it by GOD." I said, "My friend and dear brother, from my heart I will gladly see and learn this, and be thankful to my Creator for it, since it will serve and be useful toward the strengthening of my faith."

"Sit down upon the earth," he said. Then he took seven tablets and prepared them neatly according to the number of the seven planets. upon each tablet he fashioned the character or sign of its respective planet, and then he ground up the seven planets on

which the signs were found, and dropped them one
after the other, according to their special
CONSTELLATION, as was required, into a crucible,
until they melted together. He next let seven drops
of our oil fall into the crucible. Quickly there
rose a lovely, shining, flaming vapor out of the
crucible, which suffused the whole room with such
light and radiance, that I became afraid. And then I
truly saw such wonders, secrets and arcana,
including the appearance of all the planets and the
entire firmament, which revolved around the room
just as they do in heaven, that it would in no way
be proper or possible for me to describe it. I would
never have believed that such wonders were in our
Stone, if I had not seen them myself. It might
indeed be possible for a man to obtain divine
understanding through this because he can produce
such exaltation out of dead things. My peasant told
me additionally of great mysteries in regard to many
things, such as, that I could know how many true
philosophers there are in the world, who have the
Stone in our times, that I could recognize all of
them and they too recognize me, and that they would
soon get in contact with me. He taught me further
that if I should take nine drops or grains nine days
in a row that I would be gifted with angelic
understanding and believe myself to be in paradise.
through him I experienced many wonders, such that I

myself would never have believed if I had not myself seen them: sed experientia mille testes.

"Next he said, "I will show you a great and supernatural wonder, and after that I will tell you of the manifold effects, efficacy, power and virtue of our blessed Stone, and finally resolve, dispel and clarify at length all the dubious speeches, enigmas and AEQUIVOCOS SERMONES of the philosophers, through which so many people have been led astray. Lastly, I will also gladly perform certain processes which comprehend the true foundation, so that you may see that if you had understood the philosophy properly at first, then you could have attained the end in a much more rapid time. Such a failure with the MATERIA comes especially through misunderstanding of the first Resolution or dissolving and also of the correct composition, as you shall hear. For several philosophers have finished the Work and brought it to a happy conclusion in 378 days and others in thirty days."

After he had said this, we collected rainwater in a great vessel and allowed this to putrefy for some time. Then we separated the true blue water from its faces PER COHOBATIONEM and poured it in a clean, open wooden barrel, bucket or tub, and set it in the Sun. We then added a drop of OLEI NOSTRI BENEDICTI

ET INCOMBUSTIBILIS and there came successive darknesses above the abyss, just as had occurred on the first day of Creation. Next we added two drops, and the darkness soon separated from the light. Finally we added three, four, five and six drops respectively at a time, and accordingly there came forth and appeared in beauty and wonder everything that had been created in six days in the first Creation of the World, with all its details and such inexpressible gloriousness, that my senses would shatter and depart from me to tell of it; and thus it is proper for me to reveal, disclose and speak of it no more. Thus Hermes, the highly wise king, properly says in his TABULA SMARAGDINA: "Ita mundus creatus esse."

"Oh, Lord GOD," I said, "what exalted Mysteries are theses" sighing heartily and praising Him Who lives from Eternity to Eternity. At this, he said, "Dear brother, you should now be content; for I am not allowed at this time to reveal higher arcana. Pray fervently and ardently, for as soon as I am given the mandate to reveal more to you, then you shall again be instructed by me, and I shall then show you many higher things. Thus let it be for now. But let us proceed to the matters we mentioned before and examine the effects and virtues of our inexhaustible fountain, so that you may also progress properly in medicine, and thus serve and give aid to your neighbors, the poor and the sick. Therefore be seated and write it down well for much depends upon it. But first I will treat of the foundation of the three Principles, and then go on to the main point. Therefore learn this: Just as Almighty GOD alone is a single Eternal Being, through Whom all has been and is, and yet there exist in this single Being, three different persons; you should know as well that in likeness to Him it is so ordered that all things also must exist as a unity, but that in this unity there is a twofoldness: the one volatile, the other fixed and enduring; the one psychical, the other corporeal, or the one white and the other red- but the third is concealed and virtuosly placed and ordained.

"From this it follows that all things which are enduring, and which should be good and remain the same, flow AD SIMILITUDINEM out of one thing, and are divided into three, and that the three must again be compounded into one; for otherwise it would go against the desire of the Most High, and nothing of worth would come of it. The three, however, are properly body, soul and spirit or heavenly, earthly and watery, or Salt, Sulphur and Mercury, and these three are properly one and together in one thing or Subject. As in Men: body, soul and spirit; as in

GOD: Father, Son and Holy Ghost; so also in all creatures: father, mother, children. As a confirmation of this, to demonstrate His will and to show how everything should be, GOD, the Righteous and True, created Adam, His first son, in the likeness and image of Himself. Thus the individual man has been GOD'S son and image RESPECTIVE, that is, Adam and not Eve, and the whole race of man, but not in Eve. Thus from the single Adam and son of GOD have come three things: father, mother and children, and this is also to be understood of all creatures. For the earth was a wall of all four-footed animals, plants, trees, foliage and grass, and was nonetheless in the beginning a single thing, namely the earth, and then, in and together with the earth, the seed. And thus GOD created a division of the one into three when He said:

"Let the earth bring forth foliage, plants and grass and fruitful trees," which seed themselves and bear fruit according to their kind, in order, after His likeness, to increase the power of all. And thus three things have come from the single earth, namely, earth, seed, and fruit, which again bears seeds (and thus again becomes one). And although three different things have come from such a dMsion, they also again come together in that from which they have sprung; that is, all fruits go into and again become the earth, and thus again become one; as is also the case with man, because he is also taken from it in respect to his body, as GOD said: "You are earth and must again become earth."

Therefore, everything and creature goes again into that from which it has come forth, that is, into its first mother, and finally it comes, according to its situation, again to GOD, from Whom it had first gone forth by means of the Word, that is, into the MYSTERIUM MAGNUM, so that all things remain in unity and are held and endure together, namely in GOD. However, whatever departs from this and goes outside of this order of GOD is Devilish like Lucifer in his arrogance, or man in his breaking of the Commandment of GOD, or the creatures through the curse that was placed upon them because of the fall of man. However, man has again been restored to the point of becoming one and united with GOD, and of becoming GOD Himself, for a tincture or projection has occurred in Christ through the spilling of his precious blood. For he had been of a dMne nature, in that GOD had breathed a soul from His essence and being into the first Adam and into us, but he was also led astray by Satan. However, as I have said, through GOD and the man Christ something has been brought about in which the Devil may not share, because he wantonly sinned against GOD and thereby

deceived GOD'S image. GOD has permitted all of this
to occur in order to demonstrate His Almightiness
and overflowing mercy, for He wills that everything
should endure in Eternity in and according to his
ORDINATION. Therefore it is that those who undertake
or work at something against the course of Nature
and the Ordination of the Most High GOD err so
greatly in this holy Work.

"Understand me correctly," he said further. "Nature
may properly be transformed, such as when gold comes
out of silver, iron and other metals. But there must
first occur a separation and rejection of that which
should not enter into this process; that is, the
impediment, impurity or hindrance must previously be
washed away, so that the goodness that is in it may
appear freely, manifest in all its clarity. For
because of the curse placed on Nature by GOD, just
as in man, much in Nature has become corrupted;
defective and infirm. Whoever is able to remove such
deficiency and come to the aid of Nature by means of
the proper medicine (which the Artist prepares to
PLUSQUAM-PERFECTION, especially from that which is
concealed IN FOECIBUS) is a correct and true master
and philosopher. For everything bears its death and
life in itself or around its own neck; that is, its
health and sickness; and everything becomes healthy
or sick through that which is of its own kind,

127

nature and quality. To take an example from man: he is, according to his external essence, of earth or LIMBUS TERRAE SUBTILIORIS, and an extract of all earthly creatures, for which reason he is properly called the Microcosm, or the Little World. Thus the greater portion of mankind eats and drinks its sickness or health from the fruits of the earth, which is its mother. The nobler the fruits, plants and creatures are, which come from the earth and from which the man takes his nourishment, the more healthy he is and remains. And the contrary, his unhealthiness, is to be understood in the same way. Now we know that nothing in Nature is more closely related to the human body, nor has a more certain CONVENIENTIAM, than the metals, and especially the purest, namely, gold and silver, which is proven by their beautiful radiance and steadfastness. For they are able to withstand fire, which the other metals cannot do: for iron rusts, copper turns into verdigris or vitriol, and lead into quicksilver. All other things pass away in fire but gold and silver; these two endure. It is easy to conclude from this that the Spirit or Tincture enclosed in them has great steadfastness and virtue in itself, and thus has an effect on other things. Therefore, these two noble metals, which are like the human body in their nature, are able to pour out such health, when one uses them correctly and knows how to prepare them,

that the threefold point of the Universal Way could
consist of nothing but them alone. For all the
plants, roots, blossoms, etc., which easily become
corrupted, rotten and stinking, are a thousand
grades lower than these metals. However, you should
not understand all of this according to the letter,
but rather philosophically, as you were informed and
instructed by me in the beginning.

"From this finally follows that these two noble
metals, gold and silver, when restored to their
inner purity by means of the correct natural and
proper philosophical preparation are, in comparison
with the heavenly stars, like the Sun and the Moon,
which illuminate with their clarity the day and the
night, the upper and lower firmaments. And if all
the creatures were to lose the light, shine and
rediance of these, then they would die and rot, nor
could the other five planets (Mars, Saturn, Jupiter,
Venus, Mercury) and all the other fixed or unfixed
stars be fostered or preserved, no matter how
powerful they might be considered to be. In this you
can easily include the five minor metals (lead,
iron, tin, copper and quicksilver). along with all
their helpers (excluding the one which includes the
qualities of all things in itself). It is common to
all things or species that have a name or may ever
be conceived that they can do nothing and have no

effect against the health of the metals or against
the transmutation from a lesser degree to a greater
or from imperfection to perfection and purity. For
the medicine that is applied against the sickness
must always be much more noble and better than the
deficiency or corruption from which the sickness has
arisen. Therefore necessarily the cure, and transmutation of the metals cannot be sought or found in
anything but in the two Luminaries: SOLE RUBEO ET
LUNA ALBA, as the noble King Hermes says.

"For example, the first man, Adam, was born in the
presence of GOD, a living man without sin or
sickness, much less with a dead body or soul, in
which condition he would have remained, under the
Ordination and Mandate of GOD. However, when he
transgressed the Commandment, then sin and sickness
of the body and soul was engendered, so that we
poor, miserable, mortal men are now subject to
death, and to the creatures, over which we
previously had been lords and masters, so that we
are thus killed, consumed and in the end completely
eaten by our own mother, the earth, and by her
children, are actually our brothers, of our own
nature and essence. We are now just as much men in
kind, nature and quality as before the Fall, and we
shall remain men in the future, though subject to
corruption and death. But we are diminished by many

thousands of times from Perfection, and do not
appear at all like or similar to a man, when
considered against the form that the first man had
before the Fall.

"For this reason our first parents obtained through
their pleas this exalted medicine or TINCTURAM
PHILOSOPHORUM from GOD, for the preservation of a
long life and resistance to all sickness. With this
one can accomplish all the things and secrets which
I have in part revealed to you and in part must yet
conceal, until it pleases the Most High to reveal
them. However, you, an experienced man, might yet
object and ask how it is possible that the metals
should have such sympathy, correspondence, love and
friendship with men, animals and plants, since flesh
and bone and metals and minerals differ and stand as
wide apart in your eyes as heaven and earth. But
this is easy to refute if one first considers the
original Generation of men and of the metals, and
compares them one with another. For man was not, as
the uninstructed theologians and declamers think,
created by GOD the Almighty out of a simple, impure
or common lump of clay, but rather out of the best,
most subtle extraction from the very center of the
earth. Believe me, the Almighty would not use any
common earth for such a work, which He created in
His image and into which he placed, blew and

implanted a spark or ray of His eternal Essence and Being, but rather, as I have said, IPSUM EXTRACTUM TOTIUS. For if a man is resolved and separated into his three Principles, and these are again compounded, then there finally may be seen a red, lovely, radiant earth."

RELIQUA DESIDERANTUR

The End of the Lesser Edifyer.

FIGURA CABALISTICA.

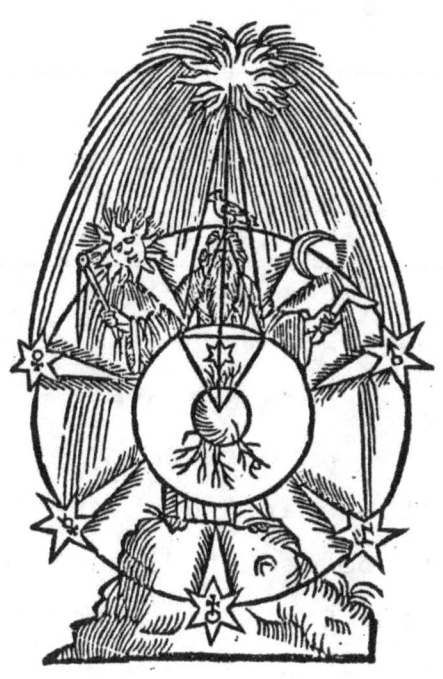

THIRD PART

PHYSICA NATURALLIS ROTUNDA
VISIONIS CHYMICAE CABALISTICA

At first the Sun and Moon appeared, with the entire firmament, and stood still, had color, but gave forth no light from themselves. Under these there stood a ball; it was of an earthy color. In its center was another little ball, which glistened snow white. When I had seen all of this, there occurred a terrifying thunderclap, which frightened me extremely. There also came a great mist, and as this slowly faded, there appeared above the Sun and Moon and the firmament a star, which shined so brightly that I could not look at it directly. In color it was redder than the Sun is wont to be. As soon as this star appeared, the entire firmament with the Sun and Moon, began to move and to frisle and leap about, while from the Star many fiery rays shot out of the firmament down upon the ball. Some of these penetrated a little, some half-way, and some completely (though this was the smallest portion) into the little ball. Those that had penetrated a little or half-way into the little ball moved the water currents, so that they began to run and all sorts of fish appeared within them. Others had the effect that the large ball turned green and brought forth all sorts of trees and fruits. And all sorts of animals and men also wandered upon it. However, the rays that had penetrated completely into the little ball moved this to such a degree, that it began to seethe, like water in a harbor, and gave off a white, clear pure steam, which then extended in the form of a star into the roots of the trees and the plants. Then there began to grow forth out of this little ball or white glistening star all sorts of trees and plants, but very slowly, which rose steadily toward the circumference of the larger ball. When they had finally come quite close, there grew out of the two ends of the ball two high, rocky mountains, which soon opened. But I saw these plants and trees steadily grow and bear to the ends of the mountains all sorts of flowers of many colors.

These then closed, and turned to buttons; red, yellow, green and white. The buttons on the mountain to the right were transparent and small, while those on the one to the left were larger, but not transparent. Rays were shooting constantly from the uppermost large star, from which more and more trees and plants grew forth out of each ball. At last there came a voice, which cried loud and clearly: "Praise be to GOD, Who created this Star, which forever shall be called the Star of Wisdom and the Eternal Light."

As I watched all this with great astonishment, I did not understand, nor could I interpret what it meant, when looking around me, I perceived the PRINCIPAL, who stood near me. He spoke to me:

"Do you understand this?" I answered, no, that I could not discover the meaning and then humbly asked if he would interpret the vision for me. "That I will tell you without difficulty," he replied, "if you will listen to me. Therefore attend it with diligence and do not be thoughtless."

"The large ball is the earth, from which all kinds of fruits grow, and through which the waters flow in order to give them moisture. But in the middle is that water which I call "corporeal" in distinction to ordinary water, of which I told you last time that it is the fertile field from which all the minerals take their origin and which receives the seeds from the heaven and the firmament. As you can see, above it is the heaven, which courses with its lights around the earth, and tells the time of the years and the days. However, if you see the Star as GOD, you are mistaken, for this Star is but a creature of GOD, though gifted with such virtues

that everything in heaven and on earth comes from
it.

"The Star, which you see, and through which the
heaven and the earth are moved, is no Star, but due
to the weakness of human nature it appears to you as
a star, for no bodily creature can see or understand
it in and of itself. It is nothing other than an
invisible fire, an eternal light, which is ordered
next to GOD, over all the heavens. It is the power,
might, form, life, virtue and preservation of all
things in heaven and on the earth, an eternal
movement, not Nature, but the Lord of Nature, a
mistress of Nature, and of all her powers and, in
sumnia, everything in everything. Therefore it is
called the Star of Wisdom and an Eternal Light.
Since it is a light in itself, not borrowed from
anything else, but rather imparted to everything,
the foundation of Wisdom is concealed within it. I
cannot tell you sufficiently what power and might is
contained in it, and even if I were able to do so,
it would be impossible for you to comprehend it, for
as little as one can fathom the Mystery of GOD and
His Majesty, even so little is it possible for you
or any other man to investigate this. However, can
it not at least be understood that there is a Lord
above it, and that it comes from Him, and that He

has thus demonstrated His Almightiness, in order that one can recognize His works?

I asked: "What is the power of this Star, and what does it mean that the other stars are driven by it, and also that it moves the waters and that the earth brings forth fruits by means of its rays?"

He answered: "Almighty GOD in all of his Works is an archetype of the future procreation, such that after he created and made the heaven and earth along with all the creatures, foliage, grass, plants, animals and men, he said: 'Be fruitful and multiply your-selves.' Behold, through these words this invisible fire had begun to dominate, to rule and to receive an impression (like that of the heaven and earth but of an attractive Astral kind). It impelled Nature to complete the work that was implanted in her and not only gave the Sun (and through it the Moon and all the stars) an enduring light, but also gave a form to the moving power and to the seed through which everything comes on and from the earth, and is thus begotten as by a bearing-mother, who attracts everything to herself, as a lodestone attracts iron; just as it had previously given form to the stars and the firmament. Note also that as the earth attracted the eternal light to itself, it also attracted the form of the firmament (which form, however, had previously been a MATERIA, namely of the Star). Therefore, if GOD the Most High had not spoken the Word, then all things would have been dead, for in a short time the creatures would have become corrupted, the created ANIMALIA would have died out, the heaven would have turned to nothingness, and thus this exalted Work (from which one can sense Almightiness and Wisdom) would have turned to rubble in a short time. And although this Fire was the first creation, living and enduring eternally and imperishable (measured against the world) it still had to be obedient to its Creator, and would have stood still and not be-friended Nature and the other creatures. But as soon as the exalted Word was spoken, then this unconsuming and vitalizing Fire coursed into the heart of Nature, and made its impression and awakened her. Only then the movement properly began as an attractive force and was implanted in the firmament and the sidereal stars, and only then the living power of procreation properly entered into all creatures.

"For this reason the Sun, Moon, stars and everything were first shown to you as dead, because they would have soon again come to an end. For they were created frozen, and frozen they would have declined, and thus the earth as well. But when this powerful Word was heard, the Sun received the seed whereby it could rejuvenate and purify itself, and likewise the Moon and all the stars, so that the earth as well could every day awaken that which had died out. Fur this reason the rays and radii shoot forth constantly anew, day and night and always, from this Star. Thus these rays are the life of the firmament, the soul, the preserver or protection from dissolution. From the Sun, Moon and stars these rays take the seed that is their highest essence, and mix within it, the living Fire which the seed that has received this changes into Nature. The seed takes the moving life into itself, and with it a great power to multiply itself and to give forth to its race uncountable numbers like itself. When this mixing has occured in the heaven or firmament, then this soul-kingdom of kindled seed separates itself off in the form of stars, and falls upon the earth, the Bearing-mother, with such vigor and force that it splits the earth (though invisible) and falls until it arrives at the center of the earth. However, in such a case, much of it remains on the surface of the earth, and part deeper in the earth, for much of it separates from the seed as an impurity, from which all sorts of plants, trees and fruits come forth and grow, each according to its kind, after its spiritual, volatile and material body has taken a nature from its star. This is the Form (as was previously mentioned). As soon as this enters into the earth, it seeks the material that is useful to it, multiplies itself in it, and is given a corporeal body, even as before it had been given a spiritual body in heaven. In this body the Fire constitutes its life by means of the great Star. Thus there appear all sorts of fruit, both large and small, for everything upon the earth was shaped previously by the Star, and thus prepared as A Form.

"That certain fruits, however may be found in one place, and not in another, is caused by the fact that the heaven is joined out of two parts, around which it revolves, as is well known to the astronomers. One part cannot enter into the other, and the stars that are in each part are not like the others, for there are as many different species as there are stars. Since one part cannot enter into the others place, it is not possible, due to the roundness of heaven and the center of the earth, that the seed of a star in one part can fall on another part and develop there, but rather it must fall where it can, and this is always straight downward.

"Thus you may infer from this that a seed that comes from a star over India may not fall into Sweden, for the star does not reach so far. Of course, it is true that a man can carry a plant or a seed to another place and plant it there, and even though its star cannot reach to that place, that it will grow equally well there, but this is due to the inexpressible (power of) Multiplication which is concealed in the spiritual Fire, and also to the fact that the seed receives its proper digestion. For again, when a plant grows well at one place and will not come forth at another, this is due to the fact that it cannot obtain its proper materials there. Among other factors, the digestion may be too weak or too strong, for a plant that grows in Arabia cannot grow in Germany, since its nature is to have a strong digestion. Likewise, a plant that grows in Germany cannot grow in Arabia, for it would be burned up there and come to nothing. Thus one now sees how the VEGETABILIA are born by means of the stars of heaven, and how each has its own influence, how they may die out just like men, and be stained and receive their VENENA through the stars. Now should we proceed further and see how man is born and ruled according to the stars?"

DE GENERATIONE HOMINIS ET ANIMALIUM

"In the case of men and animals, however, who have a sensitive life, this has another significance. When

the commingling of man and woman occurs, then the
Astra in the animal or human being rise up, and are
so strong that they master the stars, and the stars
again master the Eternal Light (for each MATERIA
attracts the Form to itself). Thus there occurs the
introjection of the stars together with the seed
into the MATRICEM. From this comes the movement, so
that the male sperm, as the efficacious part,
commingles with the female part, working within to
make a human being or an animal, according to the
form and character that the' stars and firmament had
at that time. When this body has been finished, then
comes the second introjection, that is, the astral,
material, spiritual and firmamental body with the
life.

"Now according to the constellation of the stars at
that particular time, so the human being or animal
proves to be in sensibility and thought. Through
this the diffences in men and animals may be
understood, for someone may be a man in body, but
have the disposition of a dog, a wolf or a bear, and
some animals, such as dogs, wolves and bears are
more fierce than others of the species. Thus it is
that some pious fathers may have evil sons, or
again, that an evil father may have a pious son.

"With this the Physiognomists are refuted, who wish to judge the nature of a man from his face and the form of his organs, for one man may not be like another, and yet they both have the same thoughts. For it commonly or most frequently happens that the second introjection of the stars with the soul or life into the ANIMALIA is not comparable to or is not of the same kind as the first. Thus some men may appear sad or wrathful and have a course face, yet are friendly and humble in their hearts. And again, the same thing may occur with animals, as when at times the eyes or another subtle organ like the tongue may make their disposition manifest. Therefore it is not possible from this alone sufficiently to recognize a man, though many wickedly dare to attempt it; for the same man is different in one place than another, and is more fierce or friendly at one time than at another. This is so because he is under another star and because of the type of land that he is in. For when the stars and the type of land correspond to his nature, then his disposition and the impression given to him by the heaven is strengthened; but if they are in conflict, then his disposition is distorted and repulsive and at times his body may also be afflicted. Thus one sees how the stars rule men and are either furthering or impeding to his aims. But this is said only of bestial men, who do only want the stars wish.

"But in this regard a correct man should consider that he is yet higher than all the other creatures, that is, that beyond these corporeal and firmamental bodies, he has received from Almighty GOD His breath and the third and highest body, namely an anima or psychic body, through which the two other bodies with their spirit and soul may be begotten. For in him is to be found not merely an Eternity, but rather a perpetually enduring Eternity, through which he has become the possessor not only of the earth, but also of the heaven and the entire firmament, and therefore can master and conquer this. For this reason all Impressions, Complexions and NatMties are refuted, for here the Astra should not reign, but rather the psychic body, that is, the Spirit of GOD. The deeds of men have another name among men than they should; and this is because man should rule according to the will of Him Who created him, that is, according to the will of GOD, who forbade him to do evil and commanded him to hold to the good. Through such a commandment the entire heaven is refuted, for this commandment pays no regard to the man, of whatever Complexion he might be of or to whatever he might be inclined. This commandment overturns everything. For if Almighty

145

GOD had not known that man could resist the stars, but instead had to act in accord with that to which he was inclined, then He would have given no Commandment, no have placed man as a lord over everything. If more rested on heaven than on man, then no man could be condemned on Judgement Day, but rather the blame would be placed on luck and the stars, indeed on GOD himself, far be it for me to suggest it. Thus you men do wrongly in that you say, "I am a child of Venus, a child of the Sun, a child of Mars." That is not to speak rationally, but rather bestially, that is to make idols, namely, the stars that have made and created you. Instead you should say "I am a child of GOD, the Most High, Who has given His own Son, for me, that he might redeem me and save me from the power of the Evil Spirit through his blood, bitter suffering and death." That would be the proper way to talk, that you give to Him the honor that He has deserved from you and which He is worthy of. The heathens have relied on Astronomy, have subjected themselves to the stars, and have held the same to be their gods. But you should speak and act like men, not like heathens and irrational beasts. Let this be said as a warning to you and to everyone, so that GOD should not be caused to make his punishment more severe."

I asked: "Is Astronomy then nothing, and should one reject it completely?" He answered: "Astronomy is in and of itself a glorious Art, which is to be praised and esteemed highly in so far as one uses it correctly. However, it has become greatly falsified, as I have said, because men want to make gods out of it. In the second place, the correct calculation has been lost, for the young Astronomers only make use of that which PTOLOMAEUS has described, which is improper and false, since the heavens have now become different from what he says, and have grown very much lower, slower, sluggish in their revolution and much reduced in power. He who knows the correct method uses it as a Christian and not as a heathen, that is, in order to determine the time of the day and of the year, and in part also to investigate storms, but not, as some men wish, to announce war, riot, and misfortune, and to present this as certain. No, that is false, for such things come from the will of GOD and according to the merit of men. For even though a city might have the most favorable Ascendent, it would still be possible, if the people of it led a wicked life, and acted contrary to GOD and His commandments, that punishment would not fail to appear, and GOD would send his Scourge to chastise them. You have an example of this in Sodom and Gomorrah, which neither a good nor a bad planet was able to help, for their sins themselves had Created a bad star for them, and GOD had to punish them. You may find many such examples in the Holy Scriptures; and thus a man should ask nothing concerning the stars and their signification, but rather look to GOD Himself, Who is the Lord, Who preserved Daniel and Joseph, and is yet able to preserve others.

But if a man relies on the stars and subjects himself to them, then he has not only departed from human nature and become a beast, but also has made false idols of them. If they should now show that he is Malicious, then he must become malicious, and GOD afflicts him more, so that his foolishness and ingratitude will become evident. Also, many wish to sow their plants and other things according to the Influences. This is not merely false, but is a great error, for what, in their opinion, does the plant have to do with Astronomy or the course of the heavens? Sow a plant or seed in a good fresh soil in the sunshine, and then sow the same sort of seed in an and soil which is poor and lies in shadow; thus you will see that the seed in the fresh soil comes forth sooner than the other, even though they lie close together. Although Astronomy commanded you to begin this on a certain day at a certain time, since the one has grown better than the other the one can rise and the other declines, how then do you know whether you have chosen the right time? Thus, as I have said, this doctrine is false and worthless. The ancients understood Astronomy correctly, but since then it has been greatly falsified. It is proper that every SIMPLEX should be planted and harvested according to Astronomy, but that is to be understood as follows. Everything has its own Astronomy within itself, namely its Astra, and you should pay attention to this. When a favorable Ascendent is present and its highest planet (from which its seed has come forth) is exalted, then the plant or whatever it might be is at its best and most powerful. Then a man should harvest it and use it, and pay no attention to the heaven, neither to the summer nor to the winter, but rather to the proper summer of the plant. Wait for its own heaven, for its proper autumn and the time when it has become best and grown highest through its own forces. This is the most exalted Astronomy; from the highest you can learn to recognize the lowest; by means of the time you can learn to recognize what is before your hand and when a planet is exalted. Therefore, when you know to which planet a plant is subjected, you can se~ise its exaltation by means of the Signature.

"On the other hand, something even higher follows from Astronomy. That is, when one perceives the conjunction of certain planets along with their exaltation, and at the same time also unites their GENERA (that is, metals, minerals, plants, jewels and stones) together beneath the open heaven, then the rays of the stars shoot into these bodies and augment the virtues in them, so that what might seem impossible things can be accomplished with them; not only curing, where one may heal the sicknesses of men spiritually, that is, invisibly, merely through touching, but also many wonders may be performed and accomplished through MAGIA NATURALI.

"But as to why one should do this at the time of the conjunction of the planets and stars, note this. Every new material desires a new form, and, again, every new form desires a new material. Therefore, as soon as the planets unite with one another, then this heavenly spiritual, and material desires a form which is living in everything, and because this of an attractive nature, in an instant it draws this heavenly Fire, that is, life, into itself and unites with it. Then spirit and soul are mixed with one another and united Thus the earth, which is a mother of heaven (which is a father) attracts in an instant such bodies to itself, so that these two conjunctions thus occur in that instant, from which all things of the vegetable, animal and mineral kingdoms come. When such rays fall on water or wood which is of their nature, then their virtues and spiritual bodies enter into it, through which such magic virtues may be found. Therefore, if by means of Astronomy, you know these conjunctions beforehand, prepare your metals and plants. For if these have the same star, essence, nature and complexion, then the rays will not go into the earth, but rather into these bodies (for everything loves its own likeness), through which many wonders may be performed with them PER HOC NATURAE MIRACULUM. But every philosopher and chymist should also note something else from this, of which it is not necessary to speak. SAT SAPIENTIS."

DE GENERATIONE MINERALIUM ET VEGETABILIUM

"In the third place it was mentioned that a portion of the seeds and rays given forth and projected by the Star and the firmament, fall into the center of the earth and enter into the little ball, that is, into the heart of the earth. Know now that such rays, which come and shoot so far, are purer, subtler and more spiritual than those which have not been able to attain such depths, and that, in this case, they purge themselves through the earth, as when water is purged by sand. Why? All things have been cursed by GOD, due to the falseness of the

151

first man, that is, they are as you and all men are, glutted and surrounded by filth and impurity. Thus, when such rays shoot into the earth, the coarser spirit of the seed of the firmament remains in and on the earth, from which there grow all sorts of plants A portion also falls upon the animals, from which all sorts of sickness come. But the purged rays go through the earth, like a ghost through a wall, and reach the end and the middle-point or heart of the earth, by which the earth is sustained and strengthened. For the center is more exalted than the circumference, since the circumference arises from the center, and all the power of the circumference, which in the circumference is widely attenuated, is concentrated in the center. As you can see, in the middle of man rests the soul, the spirit of the disposition, the power and the movement. Likewise, in the middle of the seed of a plant is the heavenly Fire and the number of the multiplication. The rest is only a shell and a covering for the powers.

"Thus in the center of the earth lies the corporeal water or the mineral earth (like a yolk in an egg). It is gifted with the purest and, as mentioned, with the highest, most subtle powers of the earth, for the earth take such water from it, but it takes nothing from the earth into itself, but rather multiplies itself below and rejuvenates itself like a Kingfisher. For the Word to multiply was also directed to it, and was imparted to it along with its own fire. Therefore it has its own firmament in itself, and its movement in itself, as you can see from the example of grain. For weigh one part of grain, and then at least as much, or however much more you will, of good soil and sow the grain in it. After it has finished growing, weigh each again separately, and you will find that the soil has not been diminished, but rather its perfect weight will again be found as before. Thus it has its firmament in itself, its growth and movement, etc., through which it heats, tinctures, and rejuvenates itself, and is also strengthened in its degree of multi-plication. This does not diminish, for although the metals grow out of it, they must once again die, as you see that men are born, but again that men die, and that much water runs out of the sea, but that much comes again into it. Thus, the metals likewise die, but, against this, they grow again. The ones that die leave their bodies at the place where they were engendered, but their soul or power departs from them, and is attracted by that from which it had come forth, for everything goes again to its likeness. Thus you can sense that there is no deficiency in this water, but the FAECES that it casts off from itself fall into the earth and are of

the form of the earth, from which the highest
tinctures come, All of this neither you nor any
other man has ever perceived, nor could you ever
experience or take note of it without the help of me
or of one of my companions, who are ordained to this
by GOD. Thus you may well know joy beyond other
men."

I said: "I ask you most kindly to tell me (since you
are now on the topic) why you call the little ball
'corporeal water'."

He answered: "I call it this because I can thus
present it to you in a better manner and with its
proper name, since you can also see in this figure
that it is white and glistening like mother-of-
pearl. Thus it is a water in distinction to common
earth, but a body of earth, due to its density, in
distinction to common water. These words have not
been given it by me in vain; in a little while you
shall perceive it better. But let us leave it thus
for now.

"When this pure, purged seed and mineral form enters
into the center of the earth, the life of the
mineral metals is engendered, and a movement arises
because of the unification which this immortal, heavenly spirit desires to achieve with the dead

earthly body. It pursues its likeness (which it also attracts to itself from delight and desire), until it finds it, and which it then, in the same instant mixes and unites. After sufficient digestion a tender treelet grows forth from it, shoots into the heights, and attains its branchlets, leaves, flowers and blossoms, and at last its seed, in which all the power of the entire tree resides. This seed is the end, through which one can recognize that it is ready for multiplication, just as one may see this with an ordinary plant. And no matter in what form the many plants come forth, three seeds grow from the many seeds, and± each seed or form attracts a suitable body or material to itself out of the earth according to its nature (for unquestionable the earth has the material in itself). Thus also, all sorts of heavenly seeds fall everywhere and always into the earth from the conjunctions of the stars, by means of their form and through the power of the life-creating Fire, from which all sorts of fruit are engendered. For there is no star so small, so insignificant and impotent, that it does not give forth its ANIMALIA, VEGETABILIA and MINERALIA out of itself. (Thus it is that, if the star is insignificant and weak in power, irrational men, coarse animals, bad plants and base minerals arise, and vice versa.) Thus also, just as many different sorts of plants grow near one another and are

155

harvested together, many sorts of minerals grow near one another and become known through the harvest of miners, where they are then separated from one another by means of fire.

"Know further, that if this spiritual, heavenly body is pure, clear$_1$ from the best stars and diaphanous, then it also seeks a similiar material, that is pure, clear and diaphonous, in the center of the earth. From this there grows a subtle treelet, clear and transparent, which has fine, subtle delicate branchlets that are clear as sap. On the last one the flower that had been opened closes. From it comes according to its nature, a clear transparent and pure seed. Thus are the precious jewels born, and are coloured and characterized according to their reception of the form. This you can see clearly illustrated on the mountain to your right."

I asked further: "If this is so, then it could be inferred that one may be able to find precious jewels everywhere, since the stars revolve around the whole world, but experience teaches that there are none at all to be found here, but in other places very many can be found."

He answered: "Know that these seeds fall mostly into the hot lands, and that, where the heat is strongest, there this seed (which comes only from the planets and a few other powerful stars) is sown, and purged by the great heat of the Sun, such that even before it reaches its material in the center of the earth it achieves its most pure state. And even though the sun shines as intensely in one place as in another, the cold is so great in proximity to the poles that the seed cannot be sufficiently clarified, and even if it were immediately clarified, the digestion there is still too insignificant, for there the heat usually does not remain very long. However, in the hot lands the digestion corresponds to the seed; that is, after the seed is cast from the sun or the stars, it is first purged in the SPHAERA AERIS and then likewise in the earth, which is hot, and of an attractive nature. When the seed has cast off all its impurities, it seeks and finds its pure material, according to its nature. There it grows forth with great exuberance, for the sun warms the earth through its strong, steady light, so that the best stones that can be found on earth are engendered. Thus, even though stones can also be found in Germany and in other such places, they are not like these, for all things are nobler at morning and noon than at evening and night. As you know, Arabian gold is far more excellent than Hungarian, and Hungarian more excellent than Rheinish gold; and also in regard to fruit, the oriental varities surpass the occidental in strength and virtue. It is likewise not possible that Hungarian gold could be like Arabian gold for natural reasons, for the digestion is too meagre so that in the first place, the seed cannot be sufficiently purged, and, secondly, it cannot be properly matured. Know also (as an example) that when the sun is exalted in its own house, in bright, lovely clear weather, then the seed falls in PUNCTO EXALTATIONIS. It is purged; one part remains in the air, one part in and on the earth, and one part goes to the center of the earth. This last part is also divided. The loveliest, clearest, final part, which is most highly purified, mixes with its materials, and from this grows the garnet. Another part, which is less noble, mixes with its materials and from this grows the ruby. A part that is even less noble produces the finest Arabian gold. For all of the DIAPHANACTER turn into stones. Again, one part produces sulphur of which Arabian Sulphur, and, after it, Hungarian, are held to be the best. For although many speak of other suiphurs with high praise, those who understand have judged it so.

"From the parts that have remained on and in the earth grow the best and most noble plants that one can find, among which is that plant called ALLADRUCA or ALLAKENEA, of which much could be reported.

It grows high and brings forth red and golden-yellow flowers, which are transparent and fatty, like an oil, and are well known to the Arabian peoples. That part which has remained on the earth gives forth GAMMALII in stones, wood and plants, when all of these have previously been cut off from their roots. That part which has remained in the air gives forth a growth like sheeps wool. It falls upon the earth, is sweet, and is called RUMAM, though it is unknown to you. There are also many other such things, which it is not necessary for you to know. Thus you now see the power of a single ray, and always when one planet is in conjunction or ASPECTUM TRIGONUM, etc., with another planet. Note, when its sign changes and it passes into another sign, then its rays always fall differently than before, for which reason there are so many different fruits. Furthermore, when the sun is in its exaltation with Mars, then garnets are engendered that are redder and darker in colour, rubies that are brownish and dull and gold that is reddish and impure. So is it also to be understood in regard to the other planets.

"In cold lands there grow stones, metals and plants which are not to be found in the warm lands, such as

crystal, which must have cold, for this is found
most pure, now and then, in the midnight lands.
However, its origin is not from snow (as many
pronounce) but rather from the mineral water. It is
born through Saturn in its cold Ascendent. If it
happens that Saturn is exalted in a clear heaven,
and is in conjunction with the Moon, then you obtain
the most beautiful crystals, which are clear, white
and pure. But if it is cloudy weather, and Saturn is
in conjunction with Mars, then crystals are likewise
produced, but these are dull. So is it likewise with
other planets of other colours and forms, of which I
cannot inform you sufficiently, for you would not be
able to grasp all of it. But it is thus that from
the forementioned causes sapphires, carbuncles,
UNIONEN, corals and calcedons come to us, according
to the domination and conjunction of the planets,
which are not similiar to those in the orient. Thus
you have learned how the precious jewels are
engendered.

"You should now note something well from this (since
you wish to be an experienced physician) which
cannot be more clearly described, so that the
travelling scholars, imposters and betrayers of this
Art do not learn more about it. If the seed is pure,
corporeal, and spiritual, but not fully transparent,
that which has separated from it, namely, the

crystal, is similar. But these seeds fall rarely, and there are very few of them, and thus it is that so few precious jewels are found, for the heaven does not always permit it. However, the highest power is found in such precious jewels. For among all creatures of the heavenly Fire, it is in garriets and rubies that the purest and most delicate power is to be found through a minimal preparation, mixed with the soul of the sun. The next is in the sapphire of the moon and in the emerald of Venus. If it has the nature of its likeness, it grows forth like a tree, produces its branches, that is, its veins, and spreads its blossom in the earth, which can be found in vitriol, antimony, sulphur, marcasite, talcum and cobalt. In these the blossom is pure, delicate, subtle, like sap or an extended and scattered material, and of a much more noble essence than the metal or seed that shall come from it. Similarly roses, lavender, spikenard and other aromatic plants smell much better and are lovelier while they are still in bloom, and most others also give off a much more glorious scent than when one smells the seed or distills it. Thus this blossom is much more lovely and glorious, and higher in strength, power and virtue than its metal."

I asked: "Is this blossom also called the PRIMUM ENS, of which Theophrastus described so many wonders?"

He answered: "Yes, but it is thus improperly named and thought of, for the marcasites, cobalts, etc. are not the PRIMA ENTIA, nor is that which Theophrastus taught should be distilled from them PER SUBLIMATIONEM DESTILLATIONEM, for these are rather the beginning of the seed, which may thus be called the ULTIMUM ENS SPIRITUALE METALLORUM VEL MINERALIUM. The PRIMUM ENS itself, however, lies hidden; it is the heavenly aetherial Fire, which both contains its SUBJECTUM and is united with it. You must separate these from one another through something besides Sublimation, since through Sublimation you may only obtain the flower or blossom of the metals and minerals, and thereby only pluck it from its stem and take it away from the weeds (that is, from the mountain) and from other impure things. Rather, you must first extract the PRIMUM ENS from those things in which it rests more lightly and is much more easily obtained than from its metal. For in these it is still soft, delicate, gentle, extended, volatile and pure, whereas in a metal it is compressed, hard, coarse and fixed. Similarly (as I have already mentioned) you can attain the taste and lovely scent of roses and lavender through distilling their blossoms as though through a shorter path, and your sense of smell will tell you that this is the best way to obtain it. And you can obtain this from their seeds only with difficulty, since these are compact and compressed and the ULTIMA MATERIA ROSARIUM, it is to be understood in the same way with metals. Thus in this spiritual and ULTIMO ENTE METALLORUM (which is in the process of turning itself into a seed) there is contained great power, which can easily be produced by means of the Preparation (although only with difficulty from a metal). Note this well. Thus, when the blossom has completed its time, it closes, shrinks, grows smaller, turns into a body, and a metal grows from it (after the seed has been previously provisioned spiritually by heaven). With this Nature has finished her course, for she cannot take this seed higher, and in this way the metals and minerals are engendered, as you can see in the mountain on your left."

I asked further whether this heavenly Fire was the same in the metals that are found in a pure state as in the others, for then one could extract the Fire from the minor metals just as well as from the noblest, or (which is easier) of their PRIMO ENTE, and save oneself a great expense.

He said: "I told you before, and you should have also understood it diligently, that the heaven or the heavenly firmamental Astra, whenever a new conjunction rises, draw, as an attractive material (and yet as a fire in comparision to the earth), the life from the stars, and the living Fire. The awakening of this Fire transforms itself then into the nature and form of the firmament, or the form that the firmament has attracted to itself. That is, if the spiritual body is cold, then this anima or fire will be of a cold nature. For this Fire is not subject to any Complexion; but rather it takes on its nature according to that to which it comes. In the same way other Complexions are also engendered in the other conjunctions. You should now understand from this that the life is not the same, nor likewise the nature; thus, the more noble a creature is, the more noble also is its life or the heavenly Fire within it.

"If then, the spiritual body is crude, that is, has it origin in a crude star, then it will also take a similar material to itself. From such a conjunction grow all sorts of stone, gravel, quartz, marble, grit and sand. This grit and sand is nothing but crushed stone. These stones are thought of as weeds among the mineral stones, and grow continually with the good plants, often impeding or stopping their growth. This is why it is that commonly the best mines are found in the largest mountains, for among the best plants one often finds the most weeds. Similarly, when a tree (a beech, oak or any other kind) or any other plant grows out of the earth and rises in the air, then it branches grow strong and separate the air. Such separation, however, cannot be seen by men, although it can be recognized easily by their understand[1]. The mineral plants do exactly the same. They split the earth, heave it into the heights drive it sideways and thus create mountains and valleys. For such trees are large, powerful, and strong, as is to be expected; for if such a plant is to reach from the center to the surface of the earth, it must be large and thick, and, in addition, have great power to extend itself. Now you know where the mountains come from, and what is the cause of the valleys.

"However, this mineral growth, in which all gems, minerals and stones are formed, advances very slowly, so that many believe that it does not grow at all, but rather that there is only a flow of Mercury through the Sulphuric veins, which coagulates and turns into a metal. However, this view is false and contrary to GOD, in view of His commandment to multiply. Thus Mercury is not the MATERIA of the metals, much less Sulphur. For

---

[1] This is the exact text. The author may have meant, "by their understanding." -PNW

Mercury has its origin in the planet Mercury, and is a growth in the same, for quicksilver is engendered according to the conjunction of Mercury with the other planets, in the manner mentioned previously. Mercury is the most mobile of the planets, and transforms itself into the nature of each planet with which it stands in conjunction, that is, after it has been in the houses of the Zodiac that it rules. For it is according to the planet that dominates in a conjunction that the metals are engendered, although they also take on the nature of the planets that are conjunct with it. Therefore, many different varieties of quicksilver are found; one is white, another bluish, one is gray, another blackish, one is sluggish and inert, another swift and lively. Although in itself it is a complete metal, it is often and easily changeable. For just as Mercury can swiftly change itself into the nature and qualities of the other planets, so can quicksilver easily be transmuted and made into another metal, and especially when it has arisen from a conjunction with a planet into whose metal it is to be transformed, or transmuted. Thus, if Mercury is in conjunction with the sun, and is the ruler of the conjunction, then a Mercurial tree is engendered, whose quicksilver has a spiritual solar seed within it. Therefore such quicksilver may be easily transmuted into gold. It is thus also to be understood with all the planets. However one must be experienced in order to know and recognize what each kind of quicksilver is good for, so that one may make the proper transmutation. For if one were to take the kind that has the nature of silver, or of another metal, then one must first transform it from that nature, and then make it into gold, which requires great effort. For truly the mastery of a spirit cannot proceed by means of a body, but rather this must occur by means of a spirit which is more powerful and strong than the spirit which one wants to master. Thus through luck, in your first attempt you had a Mercury that had arisen from the conjunction of Mercury, the sun and mars, in which Mercury was the ruler and dominated over the sun and mars. It was from this conjunction that the quicksilver had grown, which you precipitated and then turned into a higher metal. The effort was minimal, because Mercury had dominated over the sun and mars, and thus you could easily master it. However, if you had tried to transmute a different kind, such as one with the nature of the moon and Jupiter, you would not have been able to master it unless you had conquered it first with SPIRITU SOLIS. This you should have first considered and thus I have unveiled your error and ignorance.

"The sort of Sulphur which is also said to generate the metals, is (as was mentioned previously) not the sort that one buys and sells, but rather the kind that comes from heaven. It was called Sulphur by the ancients, because of the heavenly Fire from the stars, which dominated it. This kind is incombustible and incorruptible. No metal can be made from the other Sulphur (without the Art), for in itself it is a peculiar growth, and yet is gifted with metallic virtues, which an Artist is able to extract. The best of this kind is found in Arabia and Hungary, for they contain the excrement of rubies, carbuncles and garnets. It is as red as blood and transparent, but its preparation is truly not trivial. In itself it has more the form than the material of metals, for the materials come from the stars and are, as I have said, nothing but the excrement of the stars, and indeed, of many different species; black, green, yellow, white and brown, according to the kind and nature of the star. Thus you can now perceive that the ancients erred concerning the generation of the metals, and that no one understood it as well as Hermes.

"Since the metals grow, as I have said, and have a beginning, it follows that they must multiply themselves. But the ancients err again, in that they pronounce that gold is eternal and imperishable. No, that is false, for everything contains both life and death. Therefore, because gold has received life, it has simultaneous received death. If this were not so, then it could not be destroyed by Nature alone, nor by means of the Art. But experience has shown it to be otherwise, as you have perceived yourself. Thus gold and every other thing in the world dies, has an end, and passes away. However, as I have said, since the minerals grow slowly, they also decline slowly. For everything that grows swiftly, also declines swiftly, as is to be seen with plants, and also with Mars, Jupiter, Saturn and Venus, for they are devoured and destroyed by rust, that is, by their own embodied death. You men, however, because of your short lifetimes, cannot perceive the death of gold, and thus you have always considered it to be immortal.

"However, it can never be proven, as many of you write, that gold takes one thousand years to grow, and this is the refutation of the third error of the ancients. See, in Spain, France and Italy the fruits grow sooner than in Poland, Sweden, Denmark and Germany, and they grow even sooner in Arabia and India. Thus too, the minerals grow sooner in one place than in another, and yet each in one summer, that is, in its own summer. For (as I have said) each thing has its own summer and firmament, and thus each metal is sown in its spring, grows in its summer, brings forth fruit in its autumn, and in its winter again declines and passes away. Therefore no set time can be assigned to a particular thing, and, in regard to gold, it developes at various rates.

"Furthermore, I must reject yet another error, which is committed by several men, who deem that one metal can be transformed into another within the earth. This is false, for as little as an apple tree can grow out of a pear-tree can gold come from silver in the earth. Although certain plants can be grafted into others, what comes of this is a special variety which cannot be found in Nature, Likewise, when silver is transplanted, LASUR comes from it, but this is not gold, nor any other kind of metal, but rather a peculiar species, which nonetheless has more of the nature of silver than the qualities of the other metals. Thus, one should ignore such gross errors; for each of the metals grows by itself, and none has anything to do with the others, just as no plant has any connection with the others. Even though all of them come from the same material, namely, the earth, and the minerals all come from the corporeal water, nevertheless they do not have the same MATERIA and form, and this must be kept in mind."

I spoke to the contrary: "But it is evident, in the first place, that one can nonetheless produce gold and silver by means of the Art from ore that is poor in quality and often contains nothing or only base metals. From this it can be inferred that such poor metals and minerals may be transformed and turned into gold and silver.

"In the second place, if the metals should not be transmutable into one another, then the Art of changing lead, tin, and copper would not be possible, because the Art is an imitator of Nature, and does nor can do nothing that Nature has not already done before it. Thus Alchemy must be false and improper."

He answered: "That one can produce gold and silver from certain ores which contain neither gold nor silver, and even in which none can be found through a test, is true and possible; but this is not to be understood as you think. For (as I have said) many poor plants grow with the good plants that bear gold and silver, and are intermixed among them, such as copper, iron, tin, lead, vitriol and antimony. These plants become mature sooner than the plants of gold and silver. Thus they are also sooner corruptible. Therefore, if one harvests the Venus and Mars or the Jupiter and Saturn ores the flowers of the Sun and the Moon are often also contained or found among them, and also, at times, the seeds.

"Thus if one tests this in the ordinary way, there is sometimes a sign of gold, which means that a seed is in it. But if only the flower is there, then the result is negative, for the flower does not have the strength to withstand the powerful fire, as does the seed.

"But if one takes the ore and digests and matures it with hot, corporeal fixed species that are engendered from the conjunctions of the sun, such as antimony, vitriol, arsenic, etc. then one obtains a correct and powerful shower of blood. Since this not only matures the immature gold, but also transmutes the other immature metals, such as lead, tin, copper and iron, and turns them into gold or silver while they are still in blossom, and thus easily affected, they should be placed in an oven that is proper for them. There the flower will be matured and made enduring in a short time, although Nature would have taken a long time on them, due to her weak digestion. Thus such ores can very often now give forth their spiritual, immature gold as mature, corporeal and fixed, as I have said, and, indeed, much more of it than Nature could have perfected. Thus it is necessary to use diligence, to insure that one has the correct species for maturation, if possible, that kind that has a relation, in regard to its birth, with the sun and moon, or even better, a golden sulphur, which contains the flowers of the sun, for these are found to be very powerful and highly tinctured in such sulphuric species. Furthermore, one must be careful to digest these properly and preserve a tempered heat, which is not destructive, but rather natural and fostering of perfection; for, truly, if one wants to perfect the ore in great amounts, this requires a special diligence and understanding. Many men have had great success with small tests, but none with larger quantities, because they not only laid the ores improperly and too thickly upon one another, cut it up too much and exposed the flowers too much to the light, but also because many were of the opinion that the digestion must be more intense, and thus made the fire larger, so that in one place the ore burned, and in another experienced no fire at all. Therefore, it remained a common ore; indeed, at times more damage than utility was the result. Therefore, one must pay attention to this.

But if one ore, however volatile it may be (for many think that more can be produced from a volatile ore, than from a fixed one which is properly spoken, but not properly understood) does not contain the

flowers of the sun or of the moon, or does not
obtain them by means of Addition, then one can cook
it, roast it or fry it, however one will, and one
will still find neither gold nor silver, but rather
only that metal which is proper to it and which it
contains. Now you possess the doctrine and
instruction concerning the maturation of the ores,
in which the Art matures them just as Nature does,
and also, against and above Nature, transmutes the
metals by means of human cunning and understanding,
which Nature cannot do, for she lacks hands and
feet, and cannot bring widely separated things
together, as a man can.

In the second place, you are of the opinion that,
since no metal can transmute into another in the
earth, it should follow that such a transmutation is
also not possible through the Art. But no, a metal
that is still in the earth remains connected to its
own stem and has its own roots, its own nourishment.
As long as it remains on its stem, it can take
nothing from another metal or attract the others
nature to itself. But when it is broken off, and
thus still desires nourishment, and then obtains a
flower of the sun or the moon or some other metal
(which is commonly in the light or in the air), it
draws this eagerly to itself. This seed, which is a
spirit of the same body, attracts it into its

nature, and at times brings itself to a better state, at times to a poorer, as has also been said of the ores. However, there must be digestion present.

"But know that as soon as an ore is separated from its stem, it has departed from Nature and no longer can either grow or come to maturation without the help of men. Furthermore, if a metal is engendered in the same source as another, then they have one mother, and yet the mother is purer in one metal than in another, for the nobler the form, the nobler the material. Therefore, since there are many forms, there must be many materials.

"As the form, if it is pure, removes the excrement from the impure materials and purifies them (and only thereafter engages in a proper mixture, from which a fruit follows), so the Art also acts when one prepares a pure form and casts it upon an impure metal, which is to say that this form separates the impure form and materials from the metal and its MATERIA. When this has occurred, a correct conjunction follows, in which a pure metal is formed.

"But if the form is exalted and subtle, as I have taught you, and multiplies itself by means of itself, then it is so virtuous and powerful, that it does not need to reject the mineral excrement either from the form or from the materials of the metals that are imperfect, but rather it tinctures them all together, and renews and rejuvinates them into the highest and best metal. But, on the other hand, if one makes such a multiplying form from a lesser metal, that is, from Jupiter or Saturn, then this multiplies the imperfection and excrementa within itself so greatly, and thus becomes so powerful, that when such a form is cast upon a perfect body or metal, it imparts its own impurity so intensely that it can master even perfect bodies like gold or silver and turn them to pure lead or tin. Through this you can now perceive the power and strength of Regeneration.

"Furthermore, since gems, metals and minerals have but one origin (in regard to the mother) one can make all sorts of gems from the metals by means of the Preparation, if one has previously refined and purified them. These gems are so great in power and virtue, when the Preparation is subtle, that they are more powerful than natural gems. But, on the other hand, if one, through preparation, takes their nature from them, then one can make enduring metals out of them. For instance, one can make metal, and even gems, from every sort of vitriol, antimony, sulphur and salt, after it has been prepared. Or, on the other hand, one can make vitriol, antimony, sulphur, salt from each metal. Nevertheless there is a great difference to be found among such things, for one can find vitriol of Venus, vitriol of the Sun, of Mars, of Saturn, etc. and also antimony of Venus, antimony of the Sun, etc., and likewise with sulphur and salt. And just as there are so many different sorts of minerals, so is it the case with quicksilver, as I have mentioned already.

"Now know further that it is possible to make a salt from each sort of gem, metal and mineral, and to make a plant or a tree from the salt, though each according to its kind. Furthermore, one can make an animal and sensitive beast from this plant or tree. Thus, on the other hand, one can make a vegetable from every animal, a salt, sulphur, vitriol, etc. from every vegetable, and further, a metal from that, and finally all sorts of precious jewels. This seems to me a great thing, which few men have nonetheless experienced and perceived. You have seen one example, namely, that wine, which is a vegetable growth, has produced a salt, which is considered to be among the minerals, and furthermore, that from this salt came Saturn.

"But see, if you had thought about it, and pursued the matter properly, you could have gone further with this Saturn, and could have made the Sun, the Moon, Mars, Jupiter, Mercury, Venus, Saturn etc,. and finally all sorts of precious jewels. Or, if you had properly desired so, you could have produced a TARTAR from the salt, and made a wine, made a grape from the wine, and a vine from the grape, even to its origin. Also, if you had wished, you could have made a different plant, of whatever kind you had wished, from the Saturn by means of its transposition, then an animal of any kind, and finally a man or homunculus (though it would have no eternal soul). Now that you have heard what this Art is, and how it proceeds further, in a short time you can make vitriol of the Sun or the Moon from vitriol of Venus, and also antimony of the Sun from antimony of Saturn, and so on, so that each thing turns into another, according to your pleasure. You have now clearly learned from this what the stars may bring about through their rays, and also what this Art which man has received from GOD, is able to accomplish. I could indeed, describe it in much greater detail, but rather contemplate it yourself, and sharpen your wits thereby, for have I not told you more than all the philosophers have told you? Yet know that all things are possible for this Art, which man has received from GOD; it is far beyond

Nature. The Art does not imitate Nature, as though it were her servant, but rather it rules over Nature, perfects her, and accomplishes things which it is impossible for Nature to do. Note that man has the entire world: India, Spain, and Italy, at his disposal, and also the entire heaven. Thus, he can make in the midnight lands that which Nature is only able to make in the noon lands; and, again, that which Nature produces at sunrise, man can make at sunset, and vice versa. Thus, man can bring a plant that grows in India into Norway, which Nature cannot do, because of the firmament. But man can do it in concord with the firmament. He can prepare the materials from whatsoever he wishes to prepare and complete, from stone or wood, and prepare such a glorious growth that Nature, because of the power, virtue, form and nature of the same, will not be able to perceive it as within her own resources. "Thus man can make Arabian gold in Germany, Sweden and Poland, and Rhenish gold in Arabia. None of this is possible for Nature.

"However, this is derived from two sources. First, GOD has created all things out of one thing, and has made his Division from that so that they all have a relationship with one another. Thus it is possible for a man to restore them into one thing and to make another creature from this, like a potter, who makes whatever he wishes out of clay.

"Secondly, man has been placed by GOD as a lord over everything. For this reason knowledge has been given to him, that he may know the virtue of all things, as can be seen in the first man, who dealt with everything according to its virtue. Thus spirits have been ordained by GOD and have received wisdom from Him, whereby they may reveal this to the most excellent men, who have been chosen for it, among whom the most outstanding was Soloman, who received wisdom from GOD, which was given him through the mediation of the spirit. In the end, however, he abused it. After him the greatest has been my friend Adolphus Magnus, who is both known and unknown; after him, Hermes, and, finally, Theophrastus. After these come many others, but they were only journey-men.

I asked who this Adolphus Magnus was.

He said: "He was a man wise whose like may not be found. I will tell you something of his deeds at the end of this instruction, but now I must proceed with what is most important. I have told you previously, and also demonstrated and proven, that neither gold nor any other metal has an eternal life, but rather must die again, that is, when it has fully attained its perfection, when it has turned into or become a seed.

"When this seed has become fully mature, it sinks down again with its plant, or, as most often happens, it falls off or is harvested. If it sinks down with its plant, the stem returns to its earth and fertilizes it. The seed falls into it, dies in it, and multiplies itself. Then a new birth occurs, as with the other plants. However in 10,000 seeds it does not occur in one that it returns to its PRIMA MATERIA and to its earth, for the other plants that grow next to it commonly exclude it immediately. If it is harvested, then you have seen how men use it. But if it falls off or is broken off, then it must rot, for, due to the density of the earth, it cannot return to the MATRICAE from which it had its origin, that is, to the corporeal water. If the earth were as subtle as the air, however, the seed could fall through it and return to its origin, just as the seed that has fallen from a weed is not impeded by the air in reaching the earth. Nevertheless if this were to fall in a stony field, then it too must rot, for it cannot grow out of stone. Thus also, when, like this seed, something falls into the earth, something else may easily block it, so that it gets stuck, and rots in the course of time. So if the earth were the mother of the metals (as many fantacists, among which Aristotle was the most notable, have pronounced) then one would not have to do very much, for where ever the seed might fall, it would soon return to its origin or attain it Mercury, within which it would seed itself and grow forth anew. Their fantastic belief has a fantastic consequence, for as little as a plant can grow in the air without earth, so little can a metal grow out of the earth, or out of common quicksilver, even if one were to sow an entire wagon full. Not much needs to be said concerning this; one can observe it every day.

"It happens, however, that some rays from the stars fall and (if they are otherwise of one nature) they mix with, enter into and unite themselves with these metals, stones and minerals that have fallen off, from which GAMMAHI arise. These immediately penetrate with this influence and attain a signature from Nature. Thus many strange forms are found upon stones, wood, metals and gems, which are not noticed, although there is no form that does not have its special manifestation and virtue. One should thus pay attention to this. Know as well that no GAMMAHI are born, unless its Subject has come to an end and is dead, that is, perfected, and has been broken off from its stem, but is not yet destroyed.

For when it still is connected to its stem, it is green, has its own nurture and moisture from its roots, and thus all Astra which attempt to enter it are driven off. But when it is broken off, then it is dry, porous and desirous of receiving energy. Thus as soon as the Astra strike into it, they are retained by it, and immediately obtain its signature. However, no signature but its own can come to the green ones, for growth does not allow it.

"Thus you have heard about the dying of minerals, that these cannot attain to multiplication, for the seed rots or is crushed, is driven out by water, ground between mountains, and scattered. From this comes gold dust and pure iron and copper, all of which are seeds that have fallen off. But if the multiplication of the minerals were to proceed and be so little impeded as the multiplication of the vegetables, then one would find as many metals, jewels and varieties of minerals growing and prospering, as one finds plants. But because the multiplication is unsuccessful, the abundance of such fruits is cut short. Thus you have now heard how animals, vegetables and minerals grow and multiply themselves; you have also received a sufficient account of the Art considered in general, that when in a short time, I feel that you will transform yourself, and be silent about all of this, including what I have just said, and keep it secret, then I will explain it to you IN SPECIE, and explicate one thing after another, in order that you may know everything, and learn to recognize one thing as well as the next."

I replied: "I thank you for this exceedingly and will henceforth deal more carefully and cautiously with these things than I have before, but I also ask fervently that you also inform me what the steam and the vapour meant, which arose from the conjunction of the spiritual, corporeal and mineral in the center of the earth."

At this he looked downward, and reflected a long time, and finally said: "You already have the whole foundation of physics; wait for the proper time and you shall have the rest. You can also perceive from the words I have spoken enough to know what is yet to be understood from this." However, I did not relent, but pled as strongly as possible, until at last, after he had listened to me a long while, he said: "I rely on your promise that you will not reveal this to the unworthy, for otherwise the punishment shall apply twofold to you. Know now, that this is the highest and most excellent element of the entire figure, for by means of this we may advance from the natural to the supernatural. However, I must first give you a preliminary instruction.

"In the first place, you not only possess the comparison of the animal, vegetable and mineral kingdoms with one another, which I have imparted to you in the discussion of Transmutation, but also you may learn to recognize through this the highest and the lowest. For instance, the weakness of Nature in comparison with the Art, for it is not possible for her to have the seed, let alone to prepare it, as a man can, who, by means of the subtle operations and high understanding of men, may obtain the proper PRIMAM MATERIAM METALLORUM, and extract the heart from the center of the earth, not to speak of attaining the eternal Fire-all of which is incredible to the ignorant and the foolish.

"You have always thought that the destruction of the metals comes from the SPIRITUS VINI. No, the SPIRITUS VINI cannot do this, it is too weak and impotent to carry out such an operation, for it would be too gross that something combustible should master something incombustible. But if the SPIRITUS VINI is transformed then it is no longer SPIRITUS VINI, but rather something higher, and then it can do this. But in regard to where this transformation should come from, know that all vegetables and animals receive a particular inferior impression and Influence, from the center of the earth, and that they have concealed this within themselves, which impression is as subtle as the vision of the eye, such that a man cannot grasp it. And yet it is a fire, and one can use diligence to understand it. Therefore, note it diligently, and with great understanding. You know, as I have said, that when I have united the invisible Fire and the Life and the Movement with the sidereal stars by means of, the attractive nature of the firmament, and thus that a spiritual body has come from the aetherial fire and the astral air, then afterwards this first spiritual conjunction is able to achieve transformation, that is, to the corporeal or mineral water, from which all gems, minerals and metals have their origin. Due to the strength of this spirit, a great movement and trembling occurs (as you can see in the figure[2]), which makes the water warm and steamy, and makes the best, most subtle part (from which otherwise precious jewels would grow) spiritual. Just as when water is warmed, the best part escapes as steam and the rest, that is, the coarsest part, resolves itself again to water and falls to its origin; so here also, one part will take on something from the Form and the spiritual body will transmute itself

---

[2] No figure was included. -PNW

into the best precious stones, and its excrement into the best metals, whereas the other part, which the Form was not able to master, steals on the gentlest breezes of the earth into the heights, purges itself like water through sand, and leaves all its impurities behind in the earth. These, because of their weight, sink again in the ground to their origin, but the subtle steam does not slacken, for because of its movement, it is ever more driven, ever more impelled, and driven forth by means of the earth and natural heat until it has attained to the roots of the vegetable kingdom. And, since these are of an attractive nature even without it (for they are always suckling the earth) this spiritual steam seeps invisibly into the roots, distributes itself throughout the plant and unites with the most delicate spirit of the plants. Here it remains and is sustained by the fiery body of the spirit. One plant is better and of a more attractive nature than another, among which, I recognize the grapevine as the most excellent, for although this is not of a strongly attractive nature, it is very subtle and pure. Thus it suckles only the highest essence from the steam, and rather a lot of it in comparison with other plants, ignoring the inferior parts, which seep into the other plants. However, one part of the remaining steam that has not entered the plants goes through the earth and enters men and animals, which also possess a living attractive power, and thus rests invisibly in them.

"Now listen and open the ears of your understanding. If one takes such pure vegetables or animals (although none is to be rejected, for though the steam is purer and more abundant in one than in another, the impure can be made equal to the pure) and by means of subtle operations bring them to the point that the animal or vegetable with the mineral spirit or steam is caught in a watery body, and then separates the mineral spirit from it with great care (for in comparision with other things there is very little of it, although it has a most exalted power, brighter and clearer than day) then this spirit will bring the animal or vegetable essence to the point that it is the same as it (if you so desire). This is the foundation of the entire Art, that the spirit of the vegetable or animal should depart from its combustibility, and become imperishable or immortal. This is the key that opens all doors, and here you have the correct PRIMAM MATERIAM of gems and metals. But if I consider it correctly, this is not the, PRIMA MATERIA, but rather the threefold extraction and essence from the PRIMA MATERIA of gems. Therefore, you should praise and thank GOD the Most High in Eternity that He has considered you worthy and has given you understanding, so that you might obtain the deepest things of the earth for your use.

"Hear further concerning the PRIMA MATERIA: when it has previously been turned into a liquid, and is unlocked by means of the incombustible (not the common) vegetable spirit, then you can dissolve gold, silver, and all minerals and gems in it, and melt them like ice in warm water. You can destroy them, kill them, and make them new again; you can visibly obtain the heavenly astral spirit (as a lamp in which the eternal Fire and the power of the highest Star of Eternal Wisdom dwells) and see it, grasp it, feel and sense it, like an unconsuming fire, shining day and night beyond the radiance of the sun, moon, stars, garnets and all fire. You will perceive all the power and perfection of the entire firmament in it. May this be a great thing to you, O you creature, that you can obtain the highest thing, exalted over all the heavens and the deepest thing in the earth by means of the Art, and in a short time, whereas Nature takes long to achieve it, and, due to its subtility, cannot hold a candle to you.

"If you now unite this super-heavenly unconsuming Fire, which is higher than everything, with the most highly purged body, which is lower than everything, and make one thing from these two red and white

glistening stars (by means of the Hermaphroditic
Spirit, created from two natures, that is, from
heaven and earth, for which reason it is white and
blue, and has the correct diameter to bring the two
outermost poles together and to make a new circle
from this conjunction, which will endure through all
Eternity - and of which much more could be said),
then you will have the power of the highest and
lowest things. (Listen diligently, for such matters
are communicated very rarely). In the first place,
you have the soul from the heart of the earth; for,
just as the heart is the noblest organ in man, and
yet is only a housing and seat for the soul, so this
corporeal water is the heart of the earth and
contains all the power of the earth within it. Just
as a seed in an apple contains the whole apple and
all of its power within itself (for a tree and an
apple can again come from this seed, but from the
apple without the seed nothing can come) so is this
water full of all power and virtue and the life of
the entire earth. However, you should not take this,
but rather something higher, namely, its soul, its
spirit, not the soul from which the metals arise,
no, but rather the corporeal soul, which comes from
the body of the garnet. Here in the vegetable
kingdom you have the pure soul of the garnet; not
its body, for this cannot rise far enough, but
rather its soul, I say. But from this soul you

should take the other, indeed, the third soul, which comes about by means of subtle preparation and purgation, which you know in part, but have not yet fully brought to completion. This soul is pure, white and crystalline, such that nothing on earth is comparable to it. This is your MATERIA.

But now also hear concerning the Form. In heaven the highest planet is the sun, through which all creatures are sustained. But this sun (as I have said before) would be nothing and dead, if it were not kindled by the eternal Fire and thus were not a lamp of this, through which the other stars first receive their radiance. Now Nature takes the seed from the sun, in which the eternal light or Fire is concealed. From this seed there grow forth garnets, rubies and gold. Since you cannot obtain garnets, you should use gold. Extract the life and the seed from this and prepare it so that it is the same as a garnet. Observe now that a soul is separating from its other body. Persue this, carry out the second and third Separation, and there will no longer be a garnet, but rather something comparable to it as gold is comparable to lead; a tiny heavenly fire, pure, delicate and clear. See, this is the final Separation, and this you must still learn. Now what has become of those men who pronounce that the Art follows Nature? No, the Art is the master, and not

Nature. She may not hold her head high, but rather must be ashamed of her works, for she can never prepare this MATERIA and this Form as high as the Art can. She can never make this Fire so pure, and the center of the earth even less so. Thus Hermes has correctly written as I translate it, that this is the glory of the entire world, which strengthens the strength of the strongest, and the mastery of all subtle things. Behold, with this you may cure all vegetables, make all unfruitful trees fruitful, and turn winter to summer and summer to winter. That is, in winter you can have all the plants which are otherwise only provided by summer. Indeed, you can make a tree bear fruit five or six times in a year; you can make a good plant from a bad one, a yound fresh tree from an old rotten one, a bitter apple sweet, turn pears to cherries, and cherries again to pears, and thus transform all plants and trees into one another.

"In the second place, you can turn all imperfect metals into good ones, that is, into garnets, rubies, emeralds, pearls, etc., or to gold and silver, and indeed, into so many of these, that you are not able to express the amount. For one part will tincture not merely ten thousand parts, but rather several hundred thousand parts, and this by means of multiplication.

"In the third place, you can liberate men from all diseases, turn an old man into a young one, and make a healthy man from a sick one. You can transform the mind and thoughts of men, and make the most pious man from the wickedest knave. And whatever you might think of all of this, it is not great but rather insignificant in comparison with what follows, for the words of Hermes have not yet been sufficiently explicated.

"Listen, for now we will advance to the supernatural. This is the key to open heaven and earth, that you may enter into the highest firmament of heaven, into the center of the earth, and into the depths of the ocean. You can see through every mountain, valley, leaf, grass, animal, man, etc., and in short through everything, as though you were looking through a piece of glass. You can learn the characteristics of everything, for everything hidden will be revealed to you, everything will humble itself before you. You will be capable of everything; you will master heaven and earth; all spirits will be obedient to you, they will have to serve you and do your will. You can also come to know everything, both present and future (as much as GOD permits), which means you can create the world and receive the power of the same. However this may

193

seem, it is knowable, for it is magic and supernatural. As I have already said, when you are granted the success of attaining the completion of the natural, then you may go on to experience the supernatural. Thus you now possess what I have taught you, and, considering how poorly you have dealt with it before, guard yourself against this, and be warned."

After I had considered and contemplated all these matters, and had also listened diligently to his abuse, which had lasted a long time, it seemed to me that I stood upon a thorn or a sharp stone, and saw these visions in a deep valley. Then I heard something rustling behind me, as though someone were wearing a silken garment. Suddenly I grew afraid and looked behind me. There appeared an old ice-grey man with a long beard that reached to his belt. He was wearing a long black garment. In one hand he held a compass, in the other a carpenter's square. As he went wordlessly past me, toward the ball, he grew ever taller and larger, until the ball reached only to his belt (although it was really higher than a house) and his head reached to the Sun. Then he set his compass in the center and circumscribed the ball, so that it became perfectly round. After this, he placed his compass on the carpenter's square, and spoke. "It is one times three multiplied by itself."

Then he placed the compass again on the top of the ball and measured the distance from this to the firmament, and again from the firmament to the highest Star. He cried with a terrible voice, "This is one out of four separated by the three." Next he drew two lines from the uppermost star to the ball, such that they intersected one another, so that a triangle was formed. After this he made a quadrangle from the center with the compass, such that one corner was at the center of the ball, and a white dove sat on the opposite corner, which he called the Spirit of the Conjunctions and the VIVAFACATIONIS. But the two remaining corners were united with the two corners of the triangle. Then a flame of fire shot forth to the lines made by the triangle and remained there. Up from below there came a white glistening star, whose rays spread outwards and intermingled in the middle of the quadrangle. This star grew blood-red and shined so brightly that I could not look directly at it. It had a threefold circle or halo around it, the innermost ring golden-yellow, the second red, and the third blood—red. The light from this star grew so intense and powerful that the uppermost Star, and the Sun and the Moon of the firmament lost their radiance and turned blood—red. Likewise the earth lost its greenness, and everything turned red, for a fire shot forth from the star, which burned up the ball together with the entire firmament, so that nothing more remained, neither Sun, nor Moon, nor anything in heaven or on earth. After this the star split in two, like a mirror or a round disk, and a new ball appeared in it. This was bright and transparent and green as an emerald. Above it stood the sun, also transparent and very bright, indeed, much brighter than it had shined before. The entire firmament was there as well, but it did not revolve. Then the old man cried "Praise be to GOD, for Evil has now been suppressed, and Truth has again been revealed. Delight, you children of light; the darkness has an end. The Sun shall never set again, but rather it shall shine on you from Eternity to Eternity, and shall never more be darkened," And then he vanished.

Then the PRINCIPAL said to me: "Observe and note this figure well, which for you is the meaning of the entire Work. In this the whole Secret of Secrets resides, in natural as well as supernatural things, which are not possible for men to comprehend. But if you are pious and have faith in GOD, then you will master all of this." The PRINCIPAL spoke further: "Until now you have learned what the Art is capable of beyond Nature. Now let us see what the SUMMUM ARCANUM is, that is, the LAPIS PHILOSOPHORUM. Then you will see hereafter how far from the mark and the middle point the ordinary philosophers have shot, although they have written many large books about it, which some of them in our time often cannot understand themselves, and yet still attempt to translate into German, Therefore, may you and everyone else come here, I say, everyone; here you shall find the true foundation, and from it learn better to understand the other writings of philosophy."

I asked: "What then is the LAPIS PHILOSOPHICUS, and what is its foundation?" He answered: "The LAPIS PHILOSOPHORUM is a micrososm, created through Regeneration or rebirth, in which the perfected essence of the uppermost and the undermost stars has found its place, as in the center. For one part of it is taken from the highest vitalizing center of heaven, which is its super-heavenly light and incomprehensible fire, through which the heaven, stars, planets and all elements have their life-light, movement, power and endurance. The other part however, comes from the undermost, purest transparent center of the earth, which is a corporeal water, and imparts life, power and efficacy to the earth.

"If these two widely separated centers (from which all the powers of the world flow) are joined together through the Art by means of the spiritual

HERMAPHRODINUM or heavenly DIAMETRUM and united with one another then from them the Stone of the Wise is Compounded (as soul, body and spirit), in which the highest and lowest powers of heaven and earth are enclosed and comprehended. For this reason, because of its nature, it may be called the true, regenerated and reborn Microcosm, and, as a PLUSQUAM PERFECTUM, or more than perfect being, which rules the entire world, it may also properly be called the Lord of the Macrocosm, or the greater World. For it is an exalted and excellent Mystery of the world, whose body, soul and spirit are purely purged and regenerated ANIMAE, or rather QUINTAE ESSENTIAE, taken from the center of the hearts of the highest and lowest worlds, in consideration of which its body is an ANIMA or QUINTA ESSENTIA, as likewise its soul and spirit are, according to the nature of each, which have been purified for the third time to the highest degree and separated from their corruptibility. Therefore, its body is the center or the ANIMA EX CORDE TERRAE VEL CORPORALIS AQUAE, whereas its soul is the CENTRUM ANIMAE from the highest eternal Light, and its spirit is the CENTRUM ANIMAE from the firmamental and astral Spirit.

If one desires to attain these exalted and great powers, then the ULTIMA MATERIA LAPIDIS must be resolved IN PRIMAM and be brought through Generation to its perfection. For the MATERIA which is prepared only through common solution and coagulation does not belong to this work, because this is no Regeneration, but only a Purgation, through which the body is washed clean, to a certain degree, like a cloth, but nevertheless remains in its old being and state, still subject to corruption and impotence, Therefore another Solution is required here, indeed, a water so divine and imperishable, that it can master the Elements and rule over them. From this the vegetables and minerals are engendered, as from the PRIMA MATERIA. It enters their innermost center, seeks their life, and separates it by means of its powers from the other dead organs, putrefies it, brings it from potential to action, and then again vegetates and vivifies the dead body, and unites it with its soul and spirit, through which their life and their powers increase greatly, and finally attain their first true perfection. For only when the body is robbed of its soul by the spirit, and these two are purified and then again joined together and united, does that which before was dead become a regenerated, new and vitalizing body, which then resurrects in all clarity, snow—white and clear, no longer subject to corruption and mortality, but rather an immortal, psychical, divine and clarified body, which brings forth manifold fruit with great virtues and powers.

Owing to its causes, this Regeneration is nothing other than a separation of corruption and a restitution of imperishability, a removal of death and a restoration of life, an abolition of the elemental and a replacement of the divine powers, and finally an extirpation of Evil and an awakening of the good and advantageous being, indeed, a death and dying of the reigning unfruitful elements and the life of the repressed, immortal, divine might and strength. For, as before in the old, natural body only corruption, death and the impure CORPUS had dominion and power, and oppressed both the soul and the spirit, usurping their power; thus in the new-born body, clarified to righteousness through this power, dominion is restored and given again to the soul and the spirit, in which the Life dwells. These two illuminate and ennoble the body, and make it like them in glory, dignity, power and might to the extent that henceforth the three rule simultaneously with one another, and demonstrate and manifest their great deeds and power.

"Such is the rebirth of a new, spiritual and tempered being, an enduring, spiritual, psychical and super-heavenly might, an immortal and imperishable power, which far surpasses the old being, and is also no longer subject to Nature, but is grounded and exalted over her to the extent that, through Regeneration, new virtues and a new all-powerful eternal life will be introduced. That is, although the body was previously inert, coarse, impure, dark, corruptible, weak and impotent, through Regeneration it will become like the soul and spirit; vital, volatile, light, penetrating, pure, subtle and clear, full of power and might, immortal, incorruptible, potent and active, in order to turn imperfection to perfection and thus to sustain it. Therefore, Regeneration consists of three operations.

"First, in the killing of the body; that is, when it is resolved into PRIMAM MATERIAM and made like this, that is, vegetated, and the VIRES that sleeps in it is awakened and brought to activity; in addition to which its soul and spirit are taken from it by means of Generation. Secondly, in the purification of the body and the spirit; that is, when the external, impure, corruptible elements are removed from them, and on the other hand, the internal, invisible, hidden, incorruptible and divine elements are given to them.

"Thirdly, in the conjunction, when the pure soul and spirit are poured again into the pure body, which is thereby brought to life, and the three are united with one another, clarified, and made enduring and equally powerful. Thus Hermes has said, 'Aufer ei animam et redde el Animam', in which, along with his 'Solve et coagula' he has comprehended the whole method of the philosophical Work.

"The reason for Regeneration is that because the Lord GOD cursed the earth, that is, the elemental, corporeal and undermost part of the world, due to the terrible fall of ADAM, and Subjected it to corruption, neither the vegetable nor the mineral kingdom, nor also the animal kingdom, could again attain their original happy state and powerful nature without it, and much less could they achieve their lost perfection.

"And since the philosophical Regeneration is nothing but a purgation and separation of the good from the bad; that is, of the soul, the highest part, within which there is life, and then the spirit, and finally the body, which is dead in comparison with the other two; thus, the soul as well as the spirit and the body must have previously been separated from their corruptible nature, so that subsequently these three may become pure ANIMAE or QUINTAE ESSENTIAE in the philosophical Work. For if one gives a body (of whatever nature one will) to the regenerated ANIMAE (yet understand that the body must have a SYMBOLUM with the same ANIMA, for otherwise anything at all could come from anything at all) then the body must be in accord with the ANIMAE and not the ANIMA with the body. The reason for this is that the life in the ANIMA becomes an everenduring life when it is regenerated, and the old body in itself is dead by contrast. Therefore, the body must be in accord with the ANIMAE, and thus alive; indeed, it must become an ANIMA when judged against its old condition, so that this should not be called a body that one can see, but rather the hidden thing that is brought forth from the same, and is introduced into sublimation by means of the spirit, so that in contrast to the common body it is said to have become a corporeal spirit. Thus too, when we speak of a spirit that is regenerated, it is no longer a common spirit, but rather has become a psychical incorruptible spirit. And, in summa, the entire philosophical Work is nothing other than the creation of a new heaven and a new earth, in which the heaven is drawn downward and the earth is lifted above itself, raised into the heights and set in the place of heaven, so that this proceeds in a like manner to how the Lord GOD acted in the creation of the World.

"For in the beginning, when the Lord GOD created heaven and earth, everything was merely a water, which was the PRIMA MATERIA, in which heaven and earth with their entire host were comprehended, together with the abyss of the earth, through which they were surrounded with darkness, void and made deep without power or life. For this reason Democritus was perhaps led to constitute his ATOMOS as a watery vapor, smoke or resolved water PRO PRINCIPIIS RERUM. Above this water, which is the purest and best part, the Spirit of GOD, which is an imperishable fire and life, hovered, and preserved the same. But he forsook the abyss of the earth, which is the most unsuitable and dead part, and drew the powers of light and life only from the earth into the water, so that even in the beginning, through the Spirit of GOD, a separation occurred in the Solution and Putrefaction. Thus the Spirit of GOD surrounded only the water, that is, the uppermost, most powerful and best part, and enclosed it with His almighty power, but left the impotent and dead part lying in the abyss of darkness. Since one essence and life is comprehended in this Solution of all creatures, the Spirit of GOD, like a hen, with her chicks, properly looked after it and covered it with the wings of its almighty fruitfulness, and strengthened and matured it for the increase of its perfect light of life. Since, through the Spirit of GOD, everything was matured in this Solution and Putrefaction, and apportioned to its perfect and efficacious state of life, there followed a true essential Separation; that is, in that GOD, by means of the command of His almighty Word, first separated the most clear, most subtle, most powerful and most purified water, which surpasses the clarity and radiance of all crystals, and called this, because of its incomprehensible radiance and inexpressible clarity, the Light. But it could also be called the first water, which, among all creatures, is the perfect power and efficacy, indeed, a living source of all being, in whose flow everything bathes, sustains and refreshes itself. Therefore it is also the Form, the ACTIO and the undermost water of all things, and additionally received the first and highest place beside GOD, through whose influence the heaven, stars, planets, all elements and inferiour bodies are sustained with their powers and movements, Out of this Light there followed other Separations, in the order of their exaltation, which were ordained by GOD for the creation of the angels, the souls, and then for the preservation of the lower world. This light is also called the COELUM EMPYREUM or VIVIFICUM as well as the MUNDUS SUPERIOR VEL INVISIBILIS.

Without doubt Plato perceived his Ideas in this, since the entire visible world is comprehended by it, as by the highest creature, according to the will of GOD.

"After this there follows the second Separation, that is, the threefold sundering or the three distinctions of the upper, middle and lower water, which signify the soul, spirit and body, where all three, in a spiritual and inexpressible manner, are comprehended in the Light, or the first Water. For through the command of the Word of GOD, at each stage the most clear and subtle water goes forth from the Light or first water to its proper place, until finally the undermost, weakest and coarsest water remains in the undermost region, as the body, after which each of these waters again attains its own Separation. The uppermost water is the invisible COELUM CHRYSTALLINUM, which is counted a MATERIA in contrast to the Light, and in which as in wax, the INFLUENTIAE LUCIS IMMORTALIS ET VIVIFICAE imprint, inform and impress themselves as in a model. Thus, this is a type and impression of the Light or first fiery water, in which the ANIMAE of all creatures, that is, of all births from the Light after the angels, show forth their light and their power. The manifold IMPRESSIONES ANIMARUM result from this in the same way, for which reason it may be called the COELUM ANIMARUM, since this uppermost water is a SCATURGIO and living wellspring of all souls, which are imparted variously with their gifts and gradations to the undermost corporeal elements,

flowing into them like a spiritual, living water.
Thus, in the Hebraic language, the heaven is called
"fire-water", from ISH and MAJIM, that is, SHAMAJIM.

"The middle water is divided through this Separation
into the visible heaven, that is, into the ORBES
PLANETARUM and the firmament or PRIMUM MOBILE, from
each of which thereafter first their ESSENTIA, that
is, the seven planets and the uncountable stars, are
separated and discharged, and then again placed back
in them, where they are united and bound to their
ORBIBUS and PROPRIUS MORIBUS. This middle water, or
ORBES SYDERUM with all its stars is of the nature of
a medium, having the qualities of both the soul and
the spirit, according to the nature of the SPIRITUS
concealed in it. Therefore, when the upper and the
lower water is separated in Genesis 1, a partition
is placed between them by GOD, in order that through
this the Light or upper water could be united with
the lower corporeal water, since the soul could
never be like the body without the spirit. Thus, the
vivifying Impressions of the Light or upper water
first enter into the middle water, which is of both
a corporeal and a spiritual nature, with which the
spiritual body is suffused, which finally lets its
nature and qualities flow into the lower water, and
thus bestows upon it a harmonious body. Thus the
spirit is an invisible water concealed in the

visible, and the soul is a fiery incomprehensible water comprehended in the spirit, that is, in the invisible water.

"The lower water is corporeal and is subdivided into the four Elements, each of which is further divided into vegetables and animals. In comparison with light, fire is only a water, and thus the pure earth was also in the Solution, and was a water, but in this case a coagulated water. It was clear, diaphonous and radiantly unsullied before the Fall of ADAM, immaculate and full of power, life and soul. Even after the corruption it still has a corporeal, pure, potent water concealed in it, but in external appearance it is nothing now but a coagulated, impure, dark water. For after the Fall the lower waters, within which the uppermost, immortal powers had been deposited, and were immediately caught and permeated, were subjected to the curse and corruption, and individually afflicted with death. He, who knows correctly, how to separate and disjoin this, that is, the immortal part from the mortal and corruptible part, and how to restore the incorruptible part to its old condition and essence, imitates GOD, and has won.

1.   Three sorts of forms and materials come from
this, three sorts of activity and passivity, three
sorts of soul, spirit and body, also three sorts of
means of the upper, middle and lower, and, as a
consequence, three sorts of Separations and
Influences.

2.   Note that such purification must take place by
means of Separation and removal of the impure from
the pure, that is, when the extraneous materials and

forms and the impure elements are dissolved and taken away from the internal ESSENTIAE.

3.   Note that since the earth was in the Solution, only this should be sought, but the dark abyss, that is, the cursed earth, should be forsaken and only the living earth be purified with its spiritual water.

4.   Note that the all-highest and purest may not be united with the all-lowest and impurest without the existence of the middle.

5.   Note that, since the best and highest part of the water is spiritual, and has ventured above itself and hovers in the heights, thus our Artificial water is to be made volatile and spiritual, which can easily occur, because the Lord GOD began by introducing the PRIMAM MATERIAM of the world into water, and out of this took the purest creatures, according to their order, For in the beginning no creature was impure, because in GOD they were all good, that is, beautiful, lovely, clear, pure, useful, full of power, life, virtue and fruitfulness. Therefore, all creatures had a relationship, a flowing and outpouring, with one another; as they still have, and thus can easily be transmuted into one another.

"Therefore the Holy Ghost calls all creatures water, and fundamentally they are only water. This is also the reason that all dissolved bodies may better be mastered, resolved, purified, united with one another and restored to their first state. For without water, nothing can be purified and attain to its original perfect essence.

"However, since I have spoken of many sorts of water here, one must harmoniously understand what the correct water in our Work is, and know how to use it. The philosopher must follow the rules and regulations of the Holy Ghost in his philosophical Work, and restore each body to its proper Solution and first essence, that is, turn it into the water from which it arose and was engendered. However, this water must not be common, elemental or corruptible water, but rather the middle, fruitful, incorruptible water, the sort of water that the spirit rules, which is mutual in essence, life and powers with the upper and with the lower world, that is, which bears something of the nature and qualities not only of the Light and uppermost world, but also of the lower, elemental water. It is like a SEQUESTER, intermediary and spokesman, which is impartial and inclined to all parties, and also can adapt itself to the nature of each and take on the perfect essence of each.

"In the second place, just as this microcosmic Solution had stood for a time while the hovering Spirit of GOD matured it, in order that one part could separate from the other, thus also the philosophical Solution must be allowed to putrefy and be mastered by its warm and moist spirit, and thus be corrupted, so that the body may be made spiritual by the spirit that hovers in and over it, and so too that body, soul and spirit properly separate and be divided from one another.

"In the third place, just as in the Maturation of the water, the Lord GOD undertook the Separation and divided it into four parts, namely, into the Light and the upper, middle and lower waters, in which the entire upper and lower world is comprehended and upon which it is founded, as in four principle parts, and from which everything that lives springs forth as from a wellspring; thus also the philosopher must divide his entire Work into four parts, as the main pillars of his Artistical building, that is, into the Light, and the upper, middle and lower water, and separate or divide these from one another.

"The Light is the Form, the living water and the efficacious power and the burning radiance of the souls, or the super-heavenly incomprehensible fire.

"But the upper water is the MATERIA or the aetherial body of the souls, their vessel and seat, or the two insensitive breezes, through whose conjunction and influence a radiant, clear, crystalline superheavenly ESSENTIA, that is, soul, comes to be.

"The middle water is counted a form in contrast to the lower water, but a material in contrast to the upper water. This is the spirit, which is the body of the souls, but also the living power, form and essence of the lower corporeal water, through which this must be mastered, purified and made spiritual. For the spirit is a living water and the true AQUA VITAE, in which the upper Light lies with its crystalline water. Through it the body, that is, the lower water, is illuminated and clarified and its previously oppressed life, which was as dead, is now first properly awakened, crowned with complete power and glory, and clarified.

"The lower water is the body and the true MATERIA, in which all of the upper powers lie according to their proper measure. Thus it is a center, on which the Form has its eyes, and which it desires, and into which the invisible outpouring of all the upper waters flow, as into a lake, in which they are made enduring and remain fixed.

"But because the lower elemental water, due to the sin committed by Adam, had to depart from its first state, and came from its purity into the greatest impurity, thus all inferiour mixed bodies, together with their souls and spirits are covered and sullied to their innermost being by the curse of impurity. If these parts, the soul, the spirit and the body, are to be delivered again from impurity and restored to their original, pure and powerful state, then they must be disjoined, divided from one another, separated several times, putrified from their impurities, and thereafter each specially again restored, that is, first purified and then made like their first essence, indeed, more glorious than before. Moses calls this building and preserving paradise. Such a body then brings forth many hundredfold, nay, thousandfold fruit, for it is heavenly, spiritual and full of souls, and is nothing other than an extract of the powers of the Light and all of its various waters, indeed, an abyss full of all powers. It is a pregnant body which bears innumerable beautiful glorious children, and remains continually pregnant and thus inclined to give birth at any moment, for it has received the upper and the lower seeds without number and measure for its multiplication. Therefore it is now inclined to give and not to take, and in it heaven and earth have become one thing.

"In such a manner, then, the LAPIS PHILOSOPHORUM is an ἄναμαι φαλίοσ and compilation of everything that is in heaven and earth, but rather a full compendium of the world and an unfathomable lake, in which the upper and the lower life have poured through its channels and Influences, a regenerated microcosm, and the center which is established between the highest and the lowest, which draws the two powers into itself, as the true philosophical lodestone, which has taken the perfection of such essences into itself to illuminate and clarify the other bodies. Finally it is the band of marital duty, of the heavenly man and the earthly woman, who are bound to one another in inseparable love and gifted with incalculable fruitfulness. AMEN.

SOLI DEO GLORIA

FOURTH PART

WARNING INSTRUCTION AND DEMONSTRATION

Against all those who falsely and sinisterly persuade themselves and others, and undertake to prepare and manufacture per se and in short time the AURUM POTABILE without the process, preparation and tincture of the UNIVERSALS LAPIDIS PHILOSOPHICI.

Briefly described and published by a student of the Work of Wisdom, for the good of the Sons of the Doctrine.

PREFACE TO THE FAVORED READER

In this last age of the world and these troubled and highly difficult times, it is unfortunately found in work and deeds how severely and thirstily this divine and most precious gift, this most true universal medicine SPAGYRICAE ARTIS ET PHILOSOPHIAE ADEPTAE is despised and oppressed by the Enemy of Truth by means of his instruments through hate, envy and mockery. Yet in this alone, the MAGNUM COMPOSITUM of the upper and lower spheres, power and ability, including and comprehending the various qualities of all the elements SUB UNARIO NUMERO CABALISTICO rests in simplicity and can be found; which preserves victory and pre-eminence in all diseases and cares before all other medications, in honor and praise of GOD, Who created it; and from which alone the correct, pure, most delicious and truest AURUM POTABILE PURPURATUM has its origin and may be preserved, which is for new birth and regenerated being and was called Ambrosia and Nectar by the ancients. For the benefit of my beloved and renowned Fatherland, in which once the ADEPTA PHILOSOPHIA, which examines the secrets of nature, flourished greatly. Now it is sought again through divine destiny and PER DIVINARUM ET HUMANARUM RERUM PERSCRUTATORES to bring it forth and publish it EX PROFUNDO ABYSSI ET IMO PHILOSOPHIAE for the benefit of the true doctors and candidates, in order that the great distinction and ABUSUS between the common AURO POTABILE and AURO PHILOSOPHICO REGENERATIO may be fundamentally and truly understood and recognized, to the extent that it may be revealed EX ABDITIS NATURAE MYSTERIIS. All in praise honor and gratitude toward GOD and for the use of the present and future lovers of truth, with a warning out of Christian brotherhood that they may direct and orient their studies and labor such that they remain within the boundaries of Nature, and learn and practice within them, for outside of them no happy or fruitful success in finding the secrets and Arcana of nature and in preserving the highest end is possible. Et fugiant hoc, qui vana et sinistra persuasione et ignorantia, ex chartaceis receptiunculis, ipsisque ambigus deceptionibus idiotarum, hanc summam naturae Essentiam fabicari student, Vale.

THE FIRST CHAPTER

It is found in PHILOSOPHIA ADEPTA ET NATURAE THESAURO that there is a threefold race of gold in the things of nature. The first is astral, the second elemental, and the third metallic. Paracelsus, of high renown, applauds this sentence in his LIBELLO VEXATIONIS, saying likewise that gold is threefold: firmamental, elemental and metallic, AUREUM VELLUS says: "Triplicem esse substantiam auri, simplicem, compositam et decompositam." And there are many others of the same opinion.

However, it is reported or described by none of these men which of these three is the particularly best and most useful source from which the AURUM POTABILE MEDICORUM may be made or taken. in order to eliminate and remove this SCRUPULUM and offence,

with all its doubts, before one may advance AD PLUS
ULTRA, knowledge is necessary. If one wants to
undertake the process or preparation of this medicine, one should first be well—informed and learn
in which material and form the gold in its metallic
essence must be formed and produced before one can
make it drinkable. Thus it is found, according to
tried experience and the opinion of all
philosophers, that it is not possible to prepare a
potable medication out of gold unless the body is
reduced to its PRIMAM MATERIAM and made spiritual.
But this reduction cannot occur unless the PRIMA
PRINCIPIA AURI is found perfect and complete in such
an operation. Nam ubi haec desunt, caetera omnia
falsa et imperfecta. Thus, Petrus Bonus says: "Ubi
deficiunt vera principia, necessario deficit
progressus et generatio eorum, quae debent oriri a
principiis veris." Therefore no other means may be
used toward this undertaking and effect than a philosophical MERCURIUS ET SULPHUR EMBRIONARUM, and this
can be produced only by one well-practiced and
experienced in the manual Work, who knows well the
process of extraction, and has been sifted and
purified through the philosophical sieve, and who
knows how to separate the pure from the impure
clearly and cleanly. Then his eyes should be opened
by his practical experience, which is the mother of
knowledge, and he should come to share in occult

faith, by means of the first transmutation of obscure ignorance into light and intellectual clarity.

Through such a means, revealed by few predecessors, an upright Artist and SOPHIAE LABORIS STUDIOSUS, DIVINA OPITOLANTE GRATIA, will attain a judicious and unfailing knowledge of which gold from the above-mentioned three to take and use, which shall prove most useful to him in preparation of the AURI POTABILIS, that he may remain within the boundaries of Nature, and, exercising himself within them, shall joyfully receive a happy success in attaining AD MAJORA NATURAE ARCANA.

THE SECOND CHAPTER

If one has now found the correct subject or the true roots of potable gold, which are easy to recognize from the preceding discourse, then one should be aware that in the preparation and solution of potabile gold the body of gold should not be used, but rather the PRIMUM ENS AURI, quod in se continet verum illud medicinale Elementum ignis, omnia vincens et penctrans. So that the primordial essences and virtues of the sun are manifest in the operation with all their colors and odors, and thus may be obtained through it. And one should thus be certain, as with comforting assurance, that if the Artist advances in good spirits AD RELIQUIA, and faithfully dedicates himself to the philosophical Solution, as a qualified Son of the Doctrine in the name of GOD, then his labor and expence shall be renumerated thousandfold, and return to him again with every benefit.

It should be given here in brotherly warning that in this Christian, praiseworthy undertaking, in honor and praise toward GOD and for the comfort of the needy poor, in order that one not be deterred in the first instance by this or that impediment of error, one must enduringly pursue that which one has begun with ASSIDUIS PRECIBUS ET LABORE, and keep in mind, as Plato says "quod, in minimus rebus divinum sit implorandum auxilium; quanto plus in summis et preciosissimis rerum arcanis."

THE THIRD CHAPTER

The SOLUTIO AURI is treated by all philosophers as a great secret; indeed, as the SUMMUM ARCANUM SPAGIRICAE ARTIS, and not without reason, for there

is nothing more difficult in the sophistical Work
than to join the fixed with the volatile in the
proportional Solution and Sublimation. No one would
ever be able to learn this from reading or
philosophical study alone, but rather he must discover it himself through manual operations and the
occult faith of manifold experience in such a way
that the SOLVENS of the philosophical CLAVIS
POTENTIALIS should fruitfully fulfill its office,
and the active and passive qualities in the
penetrating and tincturing spirit of gold should
manifest themselves in a volatile and potable
nature. This may not occur before the locked gates
of the four Elements are harmoniously opened and
give forth their hidden qualities through the
breaking of the philosophical earth IN VIRGINEA
VIRTUTE. It must be intensely observed in the
Universal Solution that the SOLVENS SOLUTUM should
henceforth become an individual by means of the
proper degree of fire, and thus endures in its
spirituality and essence as well as in its
corporality and substance.

My dear and faithful Artists, out of an especially
moving affection I have communicated something in
one stroke, unmixed and pure. If you take this well
to heart with the proper fear of GOD, you shall find
in this short discourse a fuller explanation of the

philosophical secret than in many other larger books.

This work of Solution can be quickly accomplished by him who knows the correct method, but following the common process of the Particularists, gold in its resolution which is useful to the faculties of chemistry or medicine cannot be obtained. The reason is this: although it is agreed by all philosophers that the SOLUTIO AURI must proceed by means of a corrosive, without which nothing can be completely treated, such a Solution cannot be understood to apply to metallic gold, and the reason for this is that it is impossible to master gold through any corrosives. Instead it must be reduced into its PRIMAM MATERIAM by means of volatilization, so that it may again give forth its Mercury, Sulphur and Salt separately in the Solution. For Nature has completed it as highly in the composition of its body as she is able; that is, she has compounded the four Elements purely and cleanly IN DEBITO PONDERE ET MENSURE, so that it cannot be divided or consumed by any corrosive or fire, as is well known to the smelter in the gold mines or the minter of coins. For if gold could be resolved by the Art or through the addition of this or that, and thus reduced to its PRIMUM ESSE and essence, then it could also be generated again DE NOVO, that is, not fixed, unenduring and unable to withstand the tests when it is examined, since it would be volatile. For the MERCURIUS is the most powerful principal in the generation of all metals, and Nature must demonstrate her highest power upon it, so that after many long years she may be able to bring it from volatility to fixation.

However, once it has been mastered and has attained to its ULTIMAM MATERIAM, that is, has proceeded from its mineral essence to its metallic essence, it may never again be restored to a spiritual or simple nature, as little as it is possible to make milk and cream again out of butter or cheese. For in the composition of this or any metal, its Mercury is mixed with its Venus and Salt, and changed into their essences. Thus it must then remain, and nothing may be accomplished with it of use or advantage to the physician or Artist. For gold can be resolved from its metallic nature only by means of the White tincture, and only thereafter be augmented and then reduced into a medicine, However, in such a Reduction it loses all of its metallic essence, so that, once it has been dissolved, it is never again found through any tests, and may never again be restored to its metallic nature.

Therefore, if is of great consequence that one knows the office and operation of the philosophical corrosive, and even greater, from what it should be made, In this regard, it could be of great service to true and upright Artists if I could put them on the right track to find the INTROITUM AD NATURAE THESAURUM. But the ungrateful world holds me back, in view of the earnest command and high warning to philosophers not to allow such a precious jewel and pearl to fall into the possession of the unworthy. Nevertheless, for the candidates who seek after this secret EX NATURAE INSTINCTU and for the glory and praise of GOD, there be found a special instruction here, which few other authors have ever communicated so fundamentally and publically.

THE FOURTH CHAPTER

Since the materials through which the correct philosophical Solution can be made have already been mentioned, now there should follow an extensive account concerning the materials from which it is proper to make the correct corrosive, since contemporary workers stray so far from this and thus abuse it.

Through long experience and natural practice it is found that the grain of wheat does not dissolve in its generating and multiplying essence, for its body is dissolved by means of putrefaction in an externally innate humidity of the earth, through which it sprouts when it is sowed in it, and the internal humidity of the grain of wheat along with its sleeping virtues of multiplication are awakened for Generation, driven forth and brought to action.

From this it may be understood that everything associates with, delights in, and unites with that which is related to its own nature and qualities, and that in this it willingly opens and closes, and allows itself to be anatomized and dis joined by such things internally. In this alone and through nothing else it gives forth all its flowers, colors, odors, virtues and essences. One should thus be exhorted and warned through this example and parable concerning this solution of potable gold, that no species of vegetable is to be resolved with minerals, nor minerals with vegetables, but rather the metals and minerals must immediately again be broken down and dissolved by metal and mineral species of their own kind. For how many long years was this or that Artist forced to sit before the fence in this regard, not able to get across, because the HUMIDUM RADICALE remained unknown to

him, without which it is impossible to extract the
MYSTERIA ET ARCANA along with the diaphanous QUINTO
ESSE AURI, or to obtain any medications from it? For
"simile simili gaudet." Therefore, in this regard
recollect with philosophical reason and contemplate
the tested sayings of the philosophers, who say that
no water or corrosive dissolves the species of the
metals, except that which remains constant in all
examinations of species and form in respect to both
the external and the internal essence with the
MATERIA ET FORMA.

Lullius confirms this in his TESTAMENT, fol. 95:
"Quod omne metallum dissolutum indiget necessario
hospitio, quod sit de propria ipsius materia in quo
se habet nasci et multiplicari." Rupicissa says:
"Quod nulla aqua vera reductione metalla dissolvit,
nisi illa, quae in congelationibus permanent."
Albertus similarly mentions this in his tractatus,
chapter 8: "Quod ante omnia laborandum sit, ut
faciantur aquae quae habeant in se diversas
Elementorum qualitates in virtute par quas exaitent
et coagulandid, quod transmutare volunt, et qua ipsa
metalla soluta possunt congelari."

Augurellus says in Book 1: "Dissolve the gold in its
own water. With this water in times before the

ancients have brought the ship into the harbor." A caballistic author and philosopher writes that there is only one thing in the whole world that dissolves gold and silver. Paracelsus, in DE VITA LONGA, DE ELIXIRE, says: "Resolve aurum una cum substantia auri corrosivo, et id tantisper dom fiat idem cum corrosivo."

Others write that a special moisture may be found in the minerals by means of an astonishing technique, which resolves gold without force. Walburg, Augurellus, Isaacus Hollandus and others as well maintain that without the SPIRITUM VINI neither a true enduring TINCTURA SOLIS may be extracted, nor an AURUM POTABILE be made. Thus this dissolving water is called variously by many names by divers authors, such as AQUA MERCURLALIS, LUNARIA, SPIRITUS MERCURII, AQUA MINERALIS, AQUA FORTIS, AQUA VITRIOLE, AQUA REGES, ACETUM PHILOSOPHORUM, MERCURIUS MINERALIS, SAL VEGETABILE and SPIRITUS VINI.

Do not allow yourselves, dear Artists and students of Truth, to be misled by the variety of these different names, since I promised at the beginning of this fourth chapter, out of sympathy with the pious, to warn you all and inform you, although it does not seem fully proper to me, from what material this corrosive or AQUA SOLVENS should be made. Thus allow me to tell you most faithfully that it can be extracted and obtained near or next to a source of gold, for whenever this is uncovered or revealed, the other is usually also to be found not far away. If you permit this to be most highly recommended and suggested to you, then you may be assured that you will not find a similar description in very many authors, through which revelation a door will be opened for you, which you would not be able to find through many long years in the much wandering LABYBINTHO INEXTRICABILI of Alchemy, as I and others of this time have experienced.

You do not have to seek for this source of gold and that which is related to it in India, for GOD in His supreme grace has so richly blessed the Holy Roman Empire with gold ore and all sorts of mines that you will find everything that you need within it. Merely act and work so that a natural mountain rod directs you naturally or artificially, and you will come upon the right track of the metallic, elemental gold ore, and will not fail to find it. The third astral body is above us, which cannot be treated by the Artist and his Science.

Here you have in short the account and contents of the warning and instructions concerning the way and the means through which one may attain, to the knowledge and preparation for making the true and correct philosophical AURUM POTABILE and for avoiding all unnecessary expences, as well as the pure and simple truth and demonstration, brought to light from all the writings of hidden authors and from their concealed and sealed AENIGMATIBUS, so that ones needy neighbors may be served sincerely in honor, praise and gratitude towards GOD.

For the science, honor and truth of the ancient Spagyrists and renowned Chemists is not to be defended or demonstrated except through the true writings of these authors, but even more from the experience that comes from the treasure-house of the four Elements, in which the QUINTUM ESSE and the GENERAL ARCHEUS GENERALISSIMI has its kingdom. If one comes and attains to this sort of knowledge, then the ADEPTIA PHILOSOPHIA and ALCHYMIA SOPHIA LABORIS can be erected and redressed anew, as though all the books had been burned, as once occurred in the time of Diocletian in Egypt, Thus the SCIENTIA ARTIS remains in itself always and eternally unburn-able, praise be to GOD, and generates from itself through divine destiny new books and new philosophers. For GOD is the first author of all knowledge, "a quo omnis donatio bona et omne domun perfectum," Who alone has created it in its first

principles and essence, according to His eternal
goodness and incomprehensible wisdom, and has also
ordained and taught the PRACTICUM for its
preparation. The teacher and the school abide
eternally, and cannot be conquered or ruled by any
potentate. Highly praised be His eternal holy
Majesty.

THE FIFTH CHAPTER

From all of this discourse the student of Truth will
conclude that it is impossible to prepare the AURUM
POTABILE outside of the process of the Preparation
of the Universal, but rather that after the
corrosive has been used and applied, the Art
requires the Composition of the universal tincture,
of which it is said IN GENERE SUMMI PHILOSOPHI, that
it is impossible and unbelievable to extract the
QUINTAM ESSENTIAM from the Sun without corrosives,
For without a corrosive gold cannot be mastered, as
Paracelsus mentions more broadly in ARCHIDOXUS, who
is even able to say in LIBRO DE VITA LONGA that the
corrosive is greater than the gold itself. However,
in such an operation for potable gold, when the
corrosive becomes extremely sweet, without any
separation of the solvent from the solution, and the

potable gold turns volatile, "jubet hic Plato quiescere," Although I have already revealed more than too much in the preceeding, which is contained and comprehended in this "Warning, Instruction and Demonstration of how to obtain the true Potable gold", I will nevertheless instruct the pious Natural Artist, my most faithful student, who has not yet drunken from the cup of damnation, as a special NOTATU DIGNUM, in order that the tested Artist of pure humility will learn as a practiced worker and strive to discover how the firm and mighty fortress of the essence of the four Elements, along with its supreme inhabitant, may be beseiged, contested, conquered and seized, so that, in the fourth or fifth storm or attack, nothing may be withheld.

The battle-ready hero must already be well-armed for this struggle and triumph, provided with the shield of faith, with the sword of the spirit and with the helmet of salvation, for the Enemy of Truth endures with all might by means of his followers, so that out of a thousand not one is permitted the victory nor can carry off the possession that he intended to have. Therefore, set out in the name of GOD uncowed, fresh and in good spirit for the sixth, seventh and tenth storm, approaching without fear if possible, and fighting boldly and cheerfully until the commander of the seven planets, and their essence in climbing up and down the ladders, and their material pouring in and out, have fought themselves out and must surrender all together. In this victory you will not only overpower the fortress of the many Astra incorporated with their commander, but also conquer and possess the best and most beautiful city of the land in all its diaphanousness and wealth. This is the true gain, which the regenerated Elements give forth by means of such Rectification and Purgation to all legitimate sons of the Doctrine as a loyal reward of gratitude.

Therefore give praise, honor and thanks to the eternal King and Lord, the Creator of heaven and earth, through Whom alone the triumph in such a highly dangerous and difficult battle is sustained.

And do not be displeased with this parable because my pen and German discourse may be reproached as plain, straight and simple, for I am not a sophistical Spinosus, nor one highly learned. My description is not adorned with complicated, dark speeches, "quae non faciunt ad rem", for the Truth is simple and requires no costume. If from this you cannot learn the core and center of the PHILOSOPHIA ADEPTA concerning the true potable gold and its process, then you cannot be helped; for in this

insignificant tract, the precious and eternal Word
of GOD, described by the Angel in the eighth chapter
of the fourth book of Esdran, is preserved, and thus
it is demonstrated from which original PRINCIPIIS
the elemental gold and other metals are engendered
and born. Within it you may discover the true
subject of your potable gold, by which you must
abide, for, as Paracelsus says in his CHIRURGIA in
the fourth book, we should not believe anything but
that which is taught to us by the PRIMA MATERIA.

Similarly Geber in fol 4 (if I do not err) writes:
"quod aurum potabile fieri non potest, nisi per
naturam igneam ipsius quintae essentiae." However,
after this short discription of how to make gold
drinkable, one cannot so quickly proceed to the
Artificial process as the ordinary Particularists
and others attempt to persuade themselves. For time
is required before one is able properly and well to
prepare the PRINCIPIA of the bodies and the
Propriety of the spirits, which must be brought
forth with much patience and care from the innermost
recesses through many storms and battles in the
arena of force.

Although these can be perfectly and properly
obtained EX PRIMO ENTE from their metal, they
nevertheless require astonishingly subtle techniques

and administrations, with which they must be handled
and treated, so that their generative virtue is not
disrupted and driven back by an improper degree of
fire. Furthermore, it has truly been found, as
Paracelsus, that dear man; and others have observed,
that a philosopher is born from a physician, and not
the contrary. Study in this faculty without work and
experience will always demonstrate that philosophy,
must walk on stilts when alone, for manual
experience reveals and generates sharper imagination
for attaining AD ALTIORA than philosophy alone is
able to provide: "quae nihil nisi literis et
scriptis, sed haec occultara fide naturae mysteria
factis et operis indicat." Thus, for many years this
or that well-practiced worker must keep silence and
hold back the PLUS ULTRA of his concept, against his
own wishes and advantage, because it is so extremely
difficult to find the proper temperature of the
Artificially gradiated fire. If Nature could
maintain the correct measure in this, as it is possible for the practiced son of the Doctrine to do in
his Administration, then what precious things could
she not complete and accomplish with her PRINCIPIIS?
However, there are three excellent steps which she
cannot fulfill, and which he who knows Nature is
able to fulfill through divine destiny. Thus it is
of these that it is said that where Nature stops,
there the Artist should begin. It is she who

prepares the PRINCIPIA, but with imperfection and crudities, which she then hands over to the worker, for she cannot bring them to a higher state. The reasons for this are, first, that she cannot observe the gradiation of fire in Diminution and Augmentation.

Second, she cannot separate the pure from the impure. And third, she lacks the proper instruments and vessels in which the PRAEPARATIO ESSENTIALIS may occur. Therefore, through Art and Science the PRINCIPIA NATURAE may be brought out of their metallic essence and into medicinal essences with greater usefulness and virtue than it is possible for Nature to do. But, on the other hand, the Art is not able to make the various metals and minerals from her PRINCIPIIS, as Nature can.

Let me repeat in conclusion that much has been communicated here to righteous naturalists, who EX AMOBE PROXIMI investigate the mysteries and arcana of Nature, in honor and praise of GOD, so that every student of true medicine may seek to manage his faculty and vocation by creating his medications and cures EX IPSISSIMUS NATURAE FONTIBUS PURIS, which were created and ordained by GOD for this purpose, as we intended in the Beginning.

However, the PHILOPLUTI and PHILOCRIMATI, as always, instead of this, have wanted to use this divine gift for their own convenience, through which they have brought themselves as well as the gift (since it was not discovered by them and they have misused it with impostures and fraud) into great contempt, mockery and disgrace. But GOD, who is Himself the Truth, shall RANDEM and most mercifully help the pious, battle-ready, needy Artists, who have been disgraced by the others, but who have long sought the simple, ancient Truth, so that finally the lies will be separated from the Truth, the light will shine forth out of the darkness in praise and thanksgiving towards Him, and what is false will truly be perceived by the righteous. How greatly it is to be wished and prayed for that this essential, universal medicine were now IN ESSE and in use, like other, less important medicines, for it is more than necessary, if it were only GOD'S merciful will, since the manifold eclipses of the sun, according to the magical interpretation, should mean severe times of plague, and, unfortunately, due to our sins, are to be expected more and more from one year to the next, Thus is this MEDICINA SOLIS or QUINTA SOLIS ESSENTIA, which was sought so strictly and earnestly by our predecessors, held in dignity and honor, because it is, above all other essences, most closely related to the first, most perfect Being and the UNARIO NUMERO CABALISTICO.

"Hinc prima philosophia ab antiquissimis et gravissimus viris sapientia dicta, quod per eam ad altissiman rerum incorporearum et spiritualium cognitionem ascenditur, in qua et intellectus universas et robustissimus suas vires consumit. Quicunque ergo Cahos illud metaphysicum ex tenebrosa sua potestate ad actum reducere potest luminosum, hunc felicissimum finem sui intellectus obtinuisse ferunt. Quantis vero mundi fluctibus et difficultatibus, quasiatur mens et intellectus humanus in his studiosis actionibus, antequam Platonicam illam ideam ex obscuro elicere et excitare potest, ipsa expenientia et usvs docet."

I have cited this philosophy of the ancients to its end as a conclusion, so that whoever wants to institute and undertake the process for potable gold may first reflect on such philosophy and consider that it is not an ordinary labor, as several think who are frivolous and presume to undertake to prepare this quickly, for the four Elements must rather deliver this Work to us and show us how the Elements are generated. It must be investigated and observed further from what these generate themselves and by means of what they are most permanently to be

preserved under one form and species. Therefore, a
learned author writes that the AURUM POTABILE
comprises two works. The first is the preparation of
the species from which it is to be made. The second
is the preservation of this species by means of the
Astral Salt, which preserves its cloak of honor and
joins it with the band of love, so that it becomes
volatile and drinkable, and may be applied and used
further on other things.

Thus you have in short an ample, instruction
concerning what the POTABILE VEL POTABILITAS AURI
consists in, and from what it arises and is
engendered, but not according to the common opinion
of the ignorant, who deceive themselves and others
in this, saying that they can make metallic gold
drinkable and medicinal, but not knowing in the
least or being even slightly acquainted with the
philosophical process and the administration of the
Science of SOPHIAE LABORIS, in which essence and
ground of Nature the substance of gold must first be
placed and constituted before it can be forced or
reduced to potability, or can be mastered or used,
in order that it receives no damage to its Mercury,
Sulphur and Salt in the Solution, nor is weakened by
extraneous additives, nor adulterated or altered.

Therefore, dear, pious Artists, keep in mind what Petrus Bonus says: "quod longam et laboriosam experientiam requirat haec auri liquifamio vel potabilitas." For this reason do not undertake anything against her, and you shall find the MEDIUM, which lies between the ULTIMO and the PRIMO,

For thus you will remain within the boundaries and gradiations of Nature, from which the CRESCIRE ET MULTIPLICAMINI grow. Because the Word was spoken by GOD, in the first Creation, Nature must hold to her course and remain within such boundaries, outside of which no ESSENTIAE FLORES VEL COLORES AURI could ever be preserved in their potability, oiliness and liquidity. For Nature is true, as are her PRINCIPIA in the process of their maturation and preparation. But the Artist is rarely true in their administration. "Hinc illae lachrymae." Therefore, remain in the predestined essence of Nature, so that you may bring the Square of the Universal into the circumference and points of the circle, that the one Truth may give birth to an unconquerable Truth, in praise and honor toward GOD. May merciful GOD grant you and us His blessing and grace for this purpose.

THE END

FIFTH PART

THE FIRST

Everyone desires to obtain great treasures of gold and silver, precious jewels, and riches from Nature, and to become great and exalted in the eyes of the world. GOD has created all of these for this purpose, that man may use them, and be a lord over them, and thus man should recognize His divine goodness and majesty, and give Him praise, honor and thanks for these things, However, everyone wants to look for such things where GOD has simply laid them, and only on good days, without trouble, danger and labor, and be given them without having to dig or search for them or to find them. Thus, for a long time the place or path to them has been mainly unknown, and most of the piles of treasure have been hidden. For reaching them is hard and troublesome as well as dangerous, but it is not impossible to attain them. However, GOD desires that nothing of His great treasure-pile should be hidden, but rather, in this last age before the Coming of the Last Judgement, it must be revealed to the Worthy, as Christ himself says, but nevertheless very darkly so that the unworthy do not understand it, since he says that nothing must remain unrevealed. Thus we are impelled by the Spirit of GOD to announce this to the world, in accord with the Lord's will, and, indeed, it has been issued and published by us in various languages. However, either too much of this great treasure—pile is desired to be received, or else it is despised, and sought and desired from us without GOD. For men are of the opinion that one should teach them the Art (of cooking gold according to the Alchemical method) or else should run to them with great treasures, which they desire so intensely for their splendor, arrogance, war, usury, gluttony, drunkenness, concupiscence, or other sins, and share these with them. Here they might take an example from the ten virgins, where the five foolish virgins desired oil from the five wise ones, since this must be attained in a much different way, through the effort of each individual himself and in GOD. I am soon able to perceive the character of such fellows through special divine revelations, and also from their own writings, and thus our ears are covered as if by a cloud (from their bleating and crying, and only for transient gold); and they must cry in vain. Thus the many slanders and disgraces that are heard against us (which I will not further mention) shall be judged by GOD in His own time. However, since we came to recognize a while ago the diligence and earnest of both of you, which you employ in the knowledge of GOD and in the reading of the Holy Bible (although our activity was concealed from you), and which is also evidenced in your writings, we have deemed you worthy, over several thousand, of an answer, and we communicate it with the permission of GOD and the exhortation of the Holy Ghost, as follows.

There lies, in the middle of the earth or the center of the world, a mountain, which is small and large. It is soft and mild, and yet extremely rocky and hard. It is near to everything and yet invisible, in accord with divine counsel. In it the greatest treasures lie concealed, which the entire world could not pay for. However, due to the Devil's envy, which always impedes the honor of GOD and the happiness of man, it is surrounded and guarded by many fierce beasts and rapacious birds, which make the way to it, which is very difficult, also dangerous. For this reason, and because its time had not yet come, it has not been possible until now to seek for it or to find it. Nevertheless it must now be found by all worthy men, through the effort and diligence of each individual himself.

Go to this mountain on a night that is longest and darkest, and make yourselves prepared and ready through the devout prayer of your heart. Ask no man concerning the way through which the mountain may be found and encountered. Instead, follow your guide in confidence, who shall come to you and find you on the way, though you will not know him. He will bring you there at midnight, when everything is still and dark. And yet you must master yourselves with the manly disposition of a hero, in order that you not take fright and shrink away from that which you will encounter. Nevertheless, you may take no corporeal sword or any other weapon, but instead pray constantly and devoutly to GOD, and constantly repeat the words with which your guide prompts you. When you have seen the mountain, the first wonder sign that shall occur is a great and powerful wind, which will tear up the mountain exceedingly, and break apart the rocks. Lions, dragons and other terrifying beasts will also set themselves furiously and horribly before your path. But do not be afraid; stand fast; do not desire to go back and do not look back, for the wise one who leads you there will not allow any harm to come to you. Nevertheless, you have not yet discovered the treasure, although it is very near.

Soon after the wind a great earthquake will follow, which will lay everything completely flat that was left over by the wind. But do not turn away. After the earthquake, an intense fire will come, which will fully consume all the earthly materials and disclose the treasure. But you will not yet be able to see it. However, after all these events, near toward morning, it shall become very still and lovely, and you shall soon see the morning star rising into heaven, and dawn breaking across the sky. Then you shall perceive the great treasure, of which the most excellent and exalted part is an extremely exalted and excellent tincture, with which, if it were possible and pleased GOD, and the world were worthy, the entire world could be tinctured and turned into the most exalted gold. When this tincture is constantly used as your guide will instruct you, then it will make you healthy and young again, so that you will not suffer any disease in any of your bodily organs. Near this tincture you shall find all the precious jewels that could be imagined in the world, yet you should not take any
of these yourselves, but rather be content with whatever your guide gives to you. However, you must always give heartfelt thanks for this, and act very diligently so that you do not make any show of it before the world, or turn to anything that is contrary to GOD. Instead, use it well and possess it

as though you had nothing; remain frugal and secluded, and guard yourselves against all sin as diligently as possible. Otherwise, your brother and guide will turn from you and you shall again be stripped of this happiness.

For know this as good counsel: Whoever misuses this treasure and does not live as an example to the world and pure before GOD shall lose it and has little hope of ever attaining it again. When you have made yourselves ready for this Work, and feel a strong urge to perform it, then set out and do not put it off, and he, who has been asked by us to guide you, and who has been ordained to this, will meet you underway. To him you must swear an oath, which he will teach you, to act in compliance with the Fraternity, to refrain firmly from revealing anything to anyone unworthy without the consent of your Doctors, to follow him faithfully in everything, and to do whatever he teaches and says to you, yielding neither to the left nor to the right, but orienting yourselves in everything toward him. This you should and must attempt; for your prayers and heartfelt longing, toward GOD have been heard, and you and those like you have been found worthy of participation in this treasure. Be joyful; do not rely upon yourselves, but rather on your guide, and behave blamelessly towards him. For he is a worthy person, and you should do nothing without his knowing it. For, if you desire it, he shall be at your side and not forsake you, and will tell you faithfully where to meet with our convent, and also give you instruction in our rules and articles, which it is proper for you to follow, and also accompany you until that time that everything is fully revealed, the ion re-enters the Kingdom, and the course of the World is changed. O happy, worthy and beloved brothers, join with us. Thank GOD day and night for His grace; do not be over-confident; and honor your DOCTOREM and DICTATOREM. Follow him in what he shall teach you and which we cannot write here, and keep in mind where you are going, in order that he may not be grieved, and turn from you, and bring us an evil report concerning you. May GOD preserve you.

E.D.F.O.C.R. Senior

THE SECOND RESPONSE

I enter the seven circles of heaven, seize the uppermost circle with my thoughts and stand with my feet upon the lowest one. When the moonlight blinds my eyes and I stumble or fall and break a leg, then I merely make stilts, and go slowly. This is a balm for my internal Salt, which heals me again. But then I begin to sweat, and a lovely sweet water comes out of my pores, like milk and honey. Then I burn my stilts to pure ashes, so that the ash glows without smoking, and the Kings Well gives forth its marigolds. Then, three fall into the Well and produce a darkness of the World, until the moonlight breaks through once again and turns clear. The night has passed away, the sun gives forth its light; the days of the Lord approach, the heaven turns to pure fire and kindles the entire world. All four Elements melt, and a new heaven and earth are born. In these few words the treasure of the World is comprehended, in order that your desire may be fulfilled. This our brotherhood has communicated without any recompense. But if you come again into our Order on Christmas day at St. Peters, then the B.T. shall also follow you.

THE END

A Word from the Publisher

Thank you for purchasing this small work from The R.A.M.S. Library of Alchemy. During his lifetime, Hans Nintzel was dedicated to the identification, acquisition, study, retyping and, when necessary, translation of what he considered to be the most important known works on Alchemy. Hans was assisted by his sparse network of fellow Alchemists, all members of the Restorers of Alchemical Manuscripts Society (R.A.M.S.). I was an active member of R.A.M.S.

My goal is to publish all of the works originally made available through R.A.M.S. as photocopies. To facilitate this, I have chosen to have the books professionally printed. I also have a few titles that I intend to add to the original R.A.M.S. Library, selected by strict criteria established by Hans.

If you have a work on Alchemy that you believe should be a part of the R.A.M.S. Library, please contact me through R.A.M.S. Publishing Company.

Philip N. Wheeler

www.ingramcontent.com/pod-product-compliance
Lightning Source LLC
Chambersburg PA
CBHW080803180526
45168CB00006B/2308